Selenium 4
自动化测试项目实战

基于 Python 3

田春成 李靖 / 著

电子工业出版社
Publishing House of Electronics Industry
北京·BEIJING

内 容 简 介

Selenium 是目前非常流行的自动化测试工具之一。本书基于 Python 3 介绍 Selenium 4 的基本理论与操作，涉及各种高级应用，以及如何进行项目实战，并提供了详细的自动化平台部署步骤。

本书共 13 章，分为 4 篇。第 1 篇介绍了 Selenium 概况、相关的基础知识及环境的搭建步骤；第 2 篇介绍了 Selenium 涉及的各种技术，包括 Selenium 元素定位、Selenium 常用方法等；第 3 篇以大牛测试系统为例，深入探讨了如何进行项目实战与优化，详细介绍了项目重构与代码优化、数据驱动测试、Page Object 设计模式、pytest 与 Allure 报告，以及行为驱动测试等各种常用技术；第 4 篇介绍了与平台建设相关的一些实用技术，如平台的搭建与维护、项目的部署及运用 Docker 容器技术进行测试等。

为了使读者不但能掌握 Selenium 自动化测试，而且能迅速获得项目经验，本书注重理论与实践相结合，提供了大量典型的自动化测试实例，按照由浅入深、前后照应的方式来安排内容，同时提供了直播教学视频。

如果你是初学者，可以按照本书章节的先后顺序来学习，这会让你更快上手自动化测试；如果你是有经验的高级测试从业人员，可以根据自己的需求阅读本书，夯实基础，获得更多的项目设计和部署经验，以及对项目全局的认知。

未经许可，不得以任何方式复制或抄袭本书之部分或全部内容。
版权所有，侵权必究。

图书在版编目（CIP）数据

Selenium 4 自动化测试项目实战：基于 Python 3 / 田春成，李靖著. —北京：电子工业出版社，2023.6
ISBN 978-7-121-45577-3

Ⅰ. ①S… Ⅱ. ①田… ②李… Ⅲ. ①软件工具－自动检测 Ⅳ. ①TP311.561

中国国家版本馆 CIP 数据核字（2023）第 083799 号

责任编辑：董　英
印　　刷：北京天宇星印刷厂
装　　订：北京天宇星印刷厂
出版发行：电子工业出版社
　　　　　北京市海淀区万寿路 173 信箱　　邮编：100036
开　　本：787×980　1/16　印张：21.5　字数：465.4 千字
版　　次：2023 年 6 月第 1 版
印　　次：2024 年 1 月第 2 次印刷
定　　价：89.00 元

凡所购买电子工业出版社图书有缺损问题，请向购买书店调换。若书店售缺，请与本社发行部联系，联系及邮购电话：（010）88254888，88258888。

质量投诉请发邮件至 zlts@phei.com.cn，盗版侵权举报请发邮件至 dbqq@phei.com.cn。
本书咨询联系方式：faq@phei.com.cn。

推 荐 序

在这个软件技术飞速发展的时代，伴随着软件架构的不断演进，软件测试技术也随之不断完善和发展——从早年完全基于 GUI 的自动化测试，到现如今前端 GUI 自动化测试和后端 API 自动化测试并驾齐驱。从测试介入的时机来看，软件测试正在不断"左移"，即在开发的早期阶段，测试人员就会参与其中。测试人员会从软件的可测试性需求、代码质量、接口质量等多个维度来把控软件的质量。从测试分层体系的设计上看，目前很多测试都被逐渐从前端 GUI 向后端 API 或者接口迁移。

按理说，在这种情况下，前端 GUI 测试的重要性会被逐渐削弱，但事实并非如此，目前依然有大量的面向终端用户的测试用例，需要在 GUI 层面做完整的验证，而且这些 GUI 测试需求在将来很长一段时间内会长期存在，并且对于浏览器的多样性、GUI 测试的稳定性、自动化测试框架的开放性比以往任何时候都有更高的要求。为此，测试工程师非常有必要掌握扎实的主流 GUI 自动化测试技术，并且能够将其应用到实际的工程项目中。

Selenium 作为开源领域最主流的 GUI 自动化测试框架，将是你深入学习、掌握 GUI 自动化测试的不二选择。Selenium 从早期的基于 RemoteController 来规避同源策略的 1 版本，到基于 WebDriver 而大获成功的 2 版本，再到能够支持更多浏览器的 3 版本，以及最新的 4 版本，其自身在不断地完善和发展着。可以毫不夸张地说，Selenium 几乎已经成为 GUI 自动化测试事实上的行业标准。

Python 简单易学，代码精简优雅，又有大量的第三方库支持，是学习和入门自动化测试的首选开发语言。

本书系统地讲解了基于 Python 语言使用 Selenium 4 开展 GUI 自动化测试的方方面面，既有面向初级用户的基础环境搭建方法和 Selenium 基础知识，也有结合实际项目的大量工程实践。通过自我改进的重构过程，本书依次引出了可重用脚本、数据驱动、页面对象模型及 BDD 等核心概念，使读者能够循序渐进地掌握 GUI 测试的核心概念和实践方法。

此外，本书还介绍了自动化测试平台建设的基础知识，可以帮助读者拓宽视野，站在更高的层面理解自动化测试的生态体系。

纵览全书，本书内容从易到难、循序渐进，概念清晰明了，理论联系实际，知识体系全面而系统，是一本通过 Python 语言来全面掌握 Selenium 4 的好书。

茹炳晟

腾讯 Tech Lead，资深技术专家，腾讯研究院特约研究员

读者服务

微信扫码回复：45577

- 获取本书配套视频课程
- 加入"测试"读者群，与更多同道中人互动
- 获取【百场业界大咖直播合集】（持续更新），仅需 1 元

前言

笔者在 2019 年出版《Selenium 3+Python 3 自动化测试项目实战：从菜鸟到高手》一书之后，读者朋友提出了许多宝贵的意见。随着 2021 年 Selenium 4 正式发布，笔者也开始着手再版工作。3 年疫情，给每个企业、每个人都带来了诸多不便，在"内卷"越来越严重的今天，如何才能从竞争中脱颖而出呢？大家在工作之余一定要多思考、多总结，努力提高自身的技能。

Python 的语法简单且功能强大，对编程能力较弱的初学者来说，更容易学习和使用；对有编程经验的读者来说，学习 Python 的成本很低，可以在很短的时间内学会并使用 Python 来处理问题。

本书所有实例在上一版的基础之上都进行了重新设计，全部实例都可以在本地运行，方便大家快速学习。同时，本书增加了全新的项目，可以通过逐步迭代的方式快速落地自动化测试。不论有没有基础，只要按照本书介绍的路线学习，大部分读者都能在较短的时间内掌握 Web 自动化测试方法，为从事测试开发工作打下坚实的基础。

本书的初衷是提高读者的技术学习深度与广度，从而使其向测试开发工程师的道路迈进。为此，在本书的最后一篇中介绍了 Git、Docker 容器的使用方法，以及持续集成工具 Jenkins 的使用方法等。

本书最大的特点是不需要专门学习 Python，读者可以零基础入门，通过本书实例中的自动化思维，结合 Selenium 来学习 Python，并循序渐进地学会和使用 Selenium 来实现企业级项目。

本书的知识体系

本书分为 4 篇，共 13 章。

第一篇　环境篇（第 1~3 章）：主要介绍自动化测试基础知识、Selenium 的特性与发展、Selenium IDE 的使用，以及环境搭建等基础性工作。

第二篇　基础篇（第 4、5 章）：主要介绍一些 Python 基础知识、Selenium 八大定位、Selenium 常用方法等。

第三篇　项目篇（第 6~11 章）：主要介绍如何从零开始做一个自动化测试项目。首先从需求分析入手并熟悉业务流程，然后编写脚本，实现整个流程的功能，最后不断地对脚本进行重构，如函数、文件、数据驱动、Page Object 设计模式、使用 pytest 重构项目、行为驱动测试等。

第四篇　平台篇（第 12、13 章）：主要介绍自动化测试平台，包括 Git、Jenkins、Docker 容器技术及多线程测试等内容。

本书适合哪些读者

- 软件测试人员。
- 在校学生，想学习自动化测试的人员。
- 功能测试人员。
- 想深入学习自动化测试框架的人员。
- 想从事测试工作的开发人员。
- 测试经理。

本书作者

本书由田春成和李靖编写，因水平有限，错误在所难免，不当之处恳请读者批评指正，作者邮箱 2574674466@qq.com，书中配套资源及直播课程可通过微信公众号"大牛测试"获取，同时公众号中还提供了移动自动化、接口自动化、性能测试和面试宝典等专项内容。

目 录

第一篇 环境篇

第 1 章 自动化测试简介 ································· 3

1.1 什么是自动化测试 ································· 3
1.2 自动化测试的分类 ································· 4
1.3 自动化测试项目的适用条件 ························· 5
1.4 自动化测试总结 ··································· 5
1.5 为什么选择 Selenium ······························ 6
   1.5.1 Selenium 的特性 ····························· 6
   1.5.2 Selenium 的发展 ····························· 7

第 2 章 Selenium IDE 的使用 ························· 9

2.1 Selenium IDE ···································· 9
   2.1.1 Selenium IDE 的安装步骤 ··················· 10
   2.1.2 Selenium IDE 的功能界面与工具栏 ··········· 12
   2.1.3 Selenium IDE 脚本 ························· 14
   2.1.4 wait for text、assert text 和 verify text 命令 ··· 17
   2.1.5 通过实例讲解 store title 和 echo 命令 ········ 18
2.2 从 Selenium IDE 导出脚本 ······················· 18

第 3 章　Python 与 Selenium 环境搭建 ·· 21

3.1　Windows 环境下的安装 ·· 22
3.1.1　安装 Python ··· 22
3.1.2　安装 Selenium ·· 25
3.1.3　安装开发工具和 IDE ·· 28
3.1.4　搭建不同的浏览器环境 ·· 35

3.2　macOS 环境下的安装 ·· 37
3.2.1　安装 Python ··· 37
3.2.2　安装 Selenium ·· 40
3.2.3　浏览器的驱动 ·· 40

第二篇　基础篇

第 4 章　Selenium 元素定位 ·· 43

4.1　Python 基础知识 ·· 43
4.1.1　数字类型 ·· 44
4.1.2　字符串类型 ·· 44
4.1.3　常用的判断与循环语句 ·· 45
4.1.4　列表对象 ·· 47

4.2　Selenium 八大定位 ··· 58
4.2.1　id 定位 ·· 58
4.2.2　name 定位 ·· 60
4.2.3　class 定位 ·· 61
4.2.4　link_text 定位 ··· 61
4.2.5　partial_link_text 定位 ··· 63
4.2.6　CSS 定位 ··· 63
4.2.7　XPath 定位 ··· 65
4.2.8　tag_name 定位 ·· 67

4.3　表格定位 ··· 68
4.3.1　遍历表格单元格 ··· 69
4.3.2　定位表格中的特定元素 ·· 70
4.3.3　定位表格中的子元素 ··· 71

4.4　关联元素定位策略 ··· 72
4.4.1　Above 模式 ·· 73

		4.4.2	Below 模式	74
		4.4.3	Left of 模式	76
		4.4.4	Right of 模式	76
		4.4.5	Near 模式	77
		4.4.6	Chaining relative locators 模式	77

第 5 章　Selenium 常用方法　79

- 5.1　基本方法　79
- 5.2　特殊元素定位　90
  - 5.2.1　鼠标事件操作　90
  - 5.2.2　常用的键盘事件　92
  - 5.2.3　Select 操作　93
  - 5.2.4　定位一组元素　97
- 5.3　Frame 操作　99
- 5.4　上传与下载附件　101
  - 5.4.1　上传附件操作方式一　101
  - 5.4.2　上传附件操作方式二　102
  - 5.4.3　上传附件操作方式三　104
  - 5.4.4　下载附件　105
- 5.5　Cookie 操作　106
- 5.6　驱动管理模式　109
- 5.7　颜色验证　109
- 5.8　3 种等待模式　110
  - 5.8.1　强制等待模式　110
  - 5.8.2　隐式等待模式　110
  - 5.8.3　显式等待模式　111
- 5.9　多窗口切换　112
- 5.10　弹框操作　113
- 5.11　ChromeOptions　115
- 5.12　滑块操作　116
- 5.13　元素截图　117
- 5.14　JavaScript 操作页面元素　118
- 5.15　jQuery 操作页面元素　121
- 5.16　innerText 与 innerHTML　122
- 5.17　通过源码理解 By.ID　123

第三篇 项目篇

第 6 章 项目实战 ·· 129
- 6.1 项目需求分析汇总 ·· 129
  - 6.1.1 制订项目计划 ·· 130
  - 6.1.2 编写测试用例 ·· 131
- 6.2 业务场景的覆盖与分拆 ··· 133
  - 6.2.1 逐个分析页面元素 ·· 135
  - 6.2.2 分层创建脚本 ·· 142
- 6.3 项目代码总结 ·· 153

第 7 章 项目重构与代码优化 ·· 156
- 7.1 项目重构 ·· 156
  - 7.1.1 元素定位方法优化 ·· 156
  - 7.1.2 新增岗位优化 ·· 159
  - 7.1.3 代码分层优化 ·· 161
  - 7.1.4 三层架构 ·· 165
- 7.2 代码优化 ·· 168
  - 7.2.1 无人值守自动化 ··· 168
  - 7.2.2 等待时间优化 ·· 170

第 8 章 数据驱动测试 ··· 172
- 8.1 一般文件操作 ·· 173
  - 8.1.1 文本文件操作 ·· 173
  - 8.1.2 CSV 文件操作 ·· 175
  - 8.1.3 Excel 文件操作 ·· 177
  - 8.1.4 JSON 文件操作 ·· 180
  - 8.1.5 XML 文件操作 ··· 183
  - 8.1.6 YAML 文件操作 ··· 185
  - 8.1.7 文件夹操作 ··· 188
- 8.2 通过 Excel 参数，实现参数与脚本的分离 ··· 188
  - 8.2.1 创建 Excel 文件，维护测试数据 ·· 189
  - 8.2.2 Framework Log 设置 ·· 190
  - 8.2.3 初步实现数据驱动 ·· 196
- 8.3 数据驱动框架 DDT ··· 200

- 8.3.1 单元测试 200
- 8.3.2 数据驱动框架应用 210
- 8.3.3 DDT+Excel 实现循环测试 218

第 9 章 Page Object 设计模式 222
- 9.1 什么是 Page Object 222
- 9.2 Page Object 实战 223
  - 9.2.1 Common 层代码分析 224
  - 9.2.2 Base 层代码分析 229
  - 9.2.3 PageObject 层代码分析 232
  - 9.2.4 TestCases 层代码分析 235
  - 9.2.5 Data 层分析 236
  - 9.2.6 Logs 层分析 237
  - 9.2.7 Reports 层分析 238
  - 9.2.8 其他分析 239
  - 9.2.9 执行 Page Object 项目 239

第 10 章 pytest 框架实战 245
- 10.1 pytest 与 Allure 245
  - 10.1.1 pytest 的安装 246
  - 10.1.2 简单测试案例介绍 246
  - 10.1.3 引入类来管理测试方法 247
  - 10.1.4 setup 和 teardown 方法应用 248
  - 10.1.5 fixtures 功能应用 250
  - 10.1.6 pytest 如何做参数化 251
  - 10.1.7 conftest 应用 252
  - 10.1.8 运行 Selenium 255
  - 10.1.9 使用 pytest 生成测试报告 256
  - 10.1.10 集成 Allure 报告 257
- 10.2 使用 pytest 重构项目 259

第 11 章 行为驱动测试 261
- 11.1 安装环境 261
- 11.2 行为驱动之小试牛刀 262
- 11.3 基于 Selenium 的行为驱动测试 265
- 11.4 结合 Page Object 的行为驱动测试 267

第四篇 平台篇

第 12 章 测试平台维护与项目部署 ……………………………………………………… 273
- 12.1 Git 应用 ……………………………………………………… 273
  - 12.1.1 安装 Git ……………………………………………………… 274
  - 12.1.2 Git 常用操作 ……………………………………………………… 277
  - 12.1.3 运用 GitHub ……………………………………………………… 280
- 12.2 安装 Jenkins ……………………………………………………… 283
- 12.3 配置 Jenkins ……………………………………………………… 287
- 12.4 Jenkins 应用 ……………………………………………………… 292
  - 12.4.1 自由风格项目介绍 ……………………………………………………… 292
  - 12.4.2 Jenkins Pipeline ……………………………………………………… 296
- 12.5 完整的 Jenkins 自动化实例 ……………………………………………………… 304
- 12.6 项目部署 ……………………………………………………… 313
  - 12.6.1 获取当前环境模块列表 ……………………………………………………… 313
  - 12.6.2 安装项目移植所需的模块 ……………………………………………………… 315

第 13 章 Docker 容器技术与多线程测试 ……………………………………………………… 316
- 13.1 Docker 简介 ……………………………………………………… 316
- 13.2 Docker 的一般应用场景 ……………………………………………………… 318
- 13.3 Docker 的安装和简单测试 ……………………………………………………… 319
  - 13.3.1 Docker 的安装 ……………………………………………………… 319
  - 13.3.2 Docker 的简单测试 ……………………………………………………… 321
- 13.4 Python 多线程介绍 ……………………………………………………… 323
  - 13.4.1 一般方式实现多线程 ……………………………………………………… 323
  - 13.4.2 用可调用类作为参数实例化 Thread 类 ……………………………………………………… 324
  - 13.4.3 Thread 类派生子类（重写 run 方法）……………………………………………………… 325
- 13.5 使用 Docker 容器技术进行多线程测试 ……………………………………………………… 326
  - 13.5.1 Selenium Grid ……………………………………………………… 326
  - 13.5.2 安装需要的镜像 ……………………………………………………… 327
  - 13.5.3 启动 Selenium Hub ……………………………………………………… 328
  - 13.5.4 启动 Selenium Node ……………………………………………………… 328
  - 13.5.5 查看 Selenium Grid Console 界面 ……………………………………………………… 329
  - 13.5.6 Docker 环境下多线程并发执行 Selenium Grid 测试 ……………………………………………………… 330

第一篇
环 境 篇

本篇主要介绍本书所涉及的基础知识（比如自动化测试基础知识、Selenium 的特性与发展、Selenium IDE 的使用等），以及环境搭建等基础性工作。本篇对应的章节如下。

第 1 章　自动化测试简介

第 2 章　Selenium IDE 的使用

第 3 章　Python 与 Selenium 环境搭建

第 1 章
自动化测试简介

本章主要讲解自动化测试的含义、分类、项目使用,以及自动化测试工具 Selenium 的特性与发展。

1.1 什么是自动化测试

自动化测试是软件测试活动中的一个重要分支和组成部分。随着软件产业的不断发展,市场对软件开发周期的要求越来越高,于是催生了各种开发模式,如大家熟知的敏捷开发,进而对测试提出了更高的要求。此时,产生了自动化测试,即利用工具或者脚本来达到软件测试的目的,没有人工或极少人工参与的软件测试活动被称为自动化测试。自动化测试的优势如下。

- 更方便对系统进行回归测试。当软件版本发布比较频繁时，自动化测试的效果更加明显。
- 可以自动处理原本烦琐、重复的任务，提高测试的准确性和测试人员的积极性。
- 自动化测试具有重用性和一致性，即测试脚本可以在不同的版本上重复运行，且可以保证测试内容的一致性。

1.2　自动化测试的分类

维度不同，自动化测试的分类方式也不同，以下是笔者认为的比较常见的分类方式。

1. 从软件开发周期或者分层的角度来分类

（1）单元自动化测试。

单元自动化测试是指自动化地完成对代码中的类或方法进行的测试，主要关注代码实现细节及业务逻辑等方面。

（2）接口自动化测试。

接口自动化测试用于测试系统组件间接口的请求与响应。接口测试稳定性高，更适合开展自动化测试。

（3）UI自动化测试。

UI自动化测试是指用自动化技术对图形化界面进行流程和功能等方面的验证的过程。

2. 从测试目的的角度来分类

（1）功能自动化测试。

功能自动化测试主要检查实际功能是否符合用户的需求，主要以回归测试为主，涉及图形界面、数据库连接，以及其他比较稳定且不经常发生变化的元素。

（2）性能自动化测试。

性能自动化测试是依托自动化平台自动地执行性能测试、收集测试结果，并能分析测试结果的一种接近无人值守的性能测试。性能自动化测试具有以下特性。

- 为脚本创建和优化提供类库和其他模块支撑。
- 可以设定自动化任务（比如每天根据特定场景执行一轮性能测试）。
- 自动收集测试结果并存储。
- 事中监控（比如对场景执行过程中的异常错误发送自动预警邮件的功能）。
- 成熟的平台具有自动分析功能（比如自动分析哪些事务有问题、哪些资源消耗异常等）。

（3）安全自动化测试。

类似于性能自动化测试，可以将安全测试的活动自动化，比如可以定期自动扫描，进行安全预警，或发现威胁并上报。

1.3 自动化测试项目的适用条件

上线自动化测试项目需要"天时、地利、人和"，为什么这么说呢？因为对自动化测试项目的评估需要考虑各个方面，但总体来说有一些规律可循。

- 自动化测试前期投入较多，比如人力、物力、时间等。
- 软件系统界面稳定、变更少。页面变更频繁会导致代码维护成本增加。
- 项目进度压力不太大。项目时间安排比较紧迫的，不适合进行自动化测试。
- 自动化测试的脚本可以重复使用。代码重用率高可以降低开发和维护成本。
- 测试人员具备较强的编程能力。

1.4 自动化测试总结

目前，在软件测试领域，自动化测试已成趋势。越来越多的互联网公司认为，自动化测试已成为软件测试流程的重要组成部分，极大地解放了生产力。然而没有一种自动化方案可以满足 100%的需求，在评估项目及自动化模式、工具、框架设计等方面时，需要认真对待，综合考虑各种利弊得失，寻找合适的解决方案。

自动化测试最近几年的发展也很迅猛，各种工具、框架有很多，比如 Selenium、UFT、Ruby Watir 等。

自动化测试涉及一个重要名称，即"框架"。百度百科对框架的解释是："框架是一个框子（指其约束性），也是一个架子（指其支撑性）。在软件工程中，框架是整个或部分系统的可重用设计，表现为一组抽象构件及构件实例间交互的方法。同时，也可以将框架理解为可被应用开发者定制的应用骨架。"为什么很多时候要强调框架呢？主要原因如下。

- 框架的产生是为了解决某一重要问题。
- 框架有可扩展性和可集成性。可扩展性是指框架可以很容易地扩展功能和改写功能。可集成性是指可以通过暴露一些接口等方式去和其他系统进行交互。

1.5 为什么选择 Selenium

市场上的自动化测试工具有很多，选择面也比较广，笔者为什么推荐 Selenium 呢？

1.5.1 Selenium 的特性

Selenium 在自动化测试领域非常受欢迎，主要与其本身的一些特性有关系。

- Selenium 是免费开源框架。
- 支持多种浏览器，如 Chrome、Firefox、IE 等。
- 支持多种开发语言，如 Java、Python、Ruby 等，这就使得测试人员有更大的选择空间。
- 支持在多台机器上执行并发测试，可以提升自动化测试的执行效率和资源使用率。

关于工具的选择，是否开源、是否收费不应该作为评估适用性的最大权重项，而应该结合企业自身业务需求和场景做出选择。

1.5.2　Selenium 的发展

2004 年，在 ThoughtWorks 公司工作的 Jason Huggins，为了改变手工测试工作越来越繁重的状况，缔造了 Selenium 的雏形。当时仅有一套代码库，使用这套代码库可以实现页面交互操作的自动化，让手工测试人员从繁重的、重复的、附加值低的工作中解脱出来，Selenium 1 就这样诞生了。这时的 Selenium 以 JavaScript 库为后台核心，还不能脱离 JavaScript。

目前，Selenium 已经发展到 4 版本，Selenium 4 是对 Selenium 1 的继承和发展，如图 1.1 所示展示了 Selenium 框架的发展路线图。

Selenium框架的发展路线图

1版本	2版本	3版本	4版本
核心组件	核心组件	核心组件	核心组件
Selenium IDE	Selenium IDE	Selenium IDE	Selenium IDE
Selenium Grid	Selenium Grid	Selenium Grid	Selenium Grid
Selenium RC	Selenium RC	WebDriver	WebDriver
	新增组件 替换RC	去除组件	功能增强
	WebDriver	Selenium RC	添加了CDP协议，引入了Relative Locator功能等

图 1.1

其中，Selenium IDE 是 Firefox 浏览器的一个插件，依附于浏览器运行，实现了对浏览器操作的录制与回放功能，可以将录制的脚本转化为多种脚本语言（Java、Python、Ruby 等），后续章节会对此进行详细的介绍。

Selenium RC 是 Selenium 的核心组成部分，它由两个组件构成：一个是 Selenium Server 执行测试代码，也充当了 HTTP 代理服务器的角色，用于侦测、处理浏览器与应用服务器之间的 HTTP 请求通信；另一个是 Client Library，它提供了接口，用于编程语言连接 Selenium Server，主要负责发送命令给 Selenium Server，并接收测试结果。

Selenium Grid 组件的主要作用是实现并发测试，它可以实现多台测试机器和多个浏览器并发测试。每一个测试环境下的机器都被称为 Node。工作模式由一个 Hub 和若干 Node 组成。Hub 用来管理和收集 Node 的注册信息和状态信息，接受远程调用，并把请求分发给代理节点来执行。

从图 1.1 可以看到，Selenium 2 在 1 版本的基础上添加了对 WebDriver 的支持。WebDriver 提供了更简单、更便捷的 API，大幅提高了脚本代码编写效率，其原理是通过调用浏览器的 API 来定位并操作页面上的对象。此外，Selenium 2 在 1 版本的基础上扩展了很多功能，如键盘和鼠标事件、上传和下载附件等。

Selenium 和 WebDriver 原本属于两个不同的项目。关于合并 Selenium 与 WebDriver 项目的原因，WebDriver 项目的创建者 Simon Stewart 在 2009 年 8 月的邮件中写道：一部分原因是 WebDriver 弥补了 Selenium 存在的缺点（有出色的 API），另一部分原因是 Selenium 解决了 WebDriver 存在的问题（例如支持广泛的浏览器）。Selenium 4 对 3 版本主要做了如下修改。

- 新增了针对 Chrome 的 DevTools 协议，可以利用开发者工具接口做更多事情。
- 引入了相对定位器（Relative Locator）功能，方便元素定位。
- 优化了打开新窗口或新标签页的操作。
- 提供了 Capabilities 优化功能，可以更方便地对浏览器做全局设置。

通过本章的学习，读者对自动化测试的概念及如何选择工具或者框架会有初步的认识，也会对 Selenium 框架的特性、发展轨迹等有一定的了解。

第 2 章
Selenium IDE 的使用

对于 Selenium IDE，官方给出了一个总结："针对 Web 自动化的一种录制回放型的解决方案。"它提供了简单的录制流程，初学者非常容易上手。最新的 Selenium IDE 支持 Chrome 和 Firefox 浏览器。下面将对 Selenium IDE 进行详细的介绍。

2.1 Selenium IDE

相信很多初学 Selenium 的人都接触过 Selenium IDE。该工具实现了完全图形化的操作，不但支持录制，还可以将录制脚本导出生成其他编程语言的脚本（如 Java、Python 等）。

Selenium IDE 是一款基于浏览器的插件，早期版本只支持 Firefox 浏览器，最近的版本也支持 Google Chrome 浏览器，但从 3.x 版本之后，Selenium IDE 需要安装插件才可以导出脚本。下面仅介绍 Selenium IDE 的常用功能。

2.1.1　Selenium IDE 的安装步骤

Selenium IDE 的最新版本是 3.17.4（截至本书完稿之时）。Firefox 版本的 Selenium IDE 的安装步骤如下。

（1）首先打开 Selenium IDE 的官方网站，如图 2.1 所示。

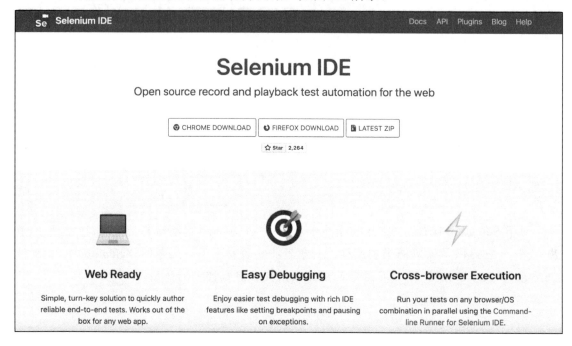

图 2.1

（2）单击"FIREFOX DOWNLOAD"按钮，进入 Selenium IDE 插件页面，如图 2.2 所示。然后单击"Download Firefox and get the extension"按钮，可以下载最新版本的 Firefox 浏览器安装包，版本号为 110.0（截至本书完稿之时）。单击按钮下面的"Download file"超链接可以下载最新版本的 Selenium IDE 插件安装文件，文件名为 selenium_ide-3.17.4.xpi，其版本号为 3.17.4（截至本书完稿之时）。

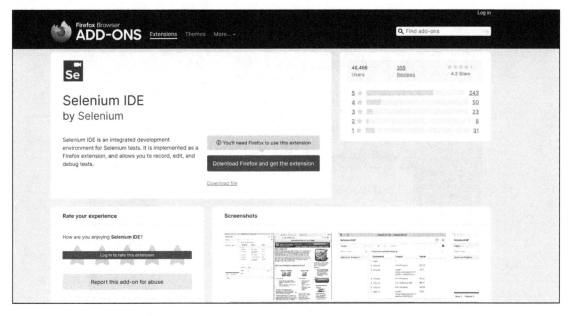

图 2.2

（3）Firefox 与 Selenium IDE 插件安装完毕后，可以打开浏览器，在其搜索栏右侧可以发现 Selenium IDE 的图标，具体如图 2.3 所示。

图 2.3

打开 Selenium IDE，其图形化界面如图 2.4 所示。

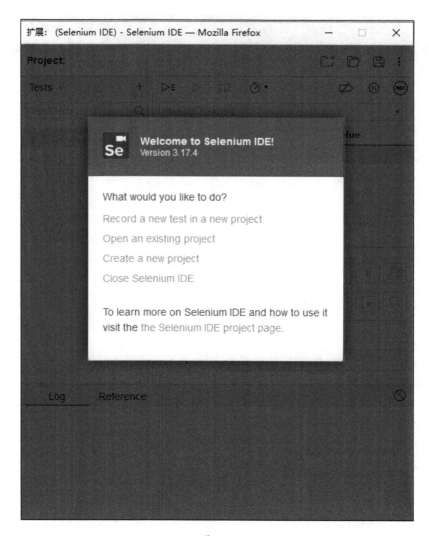

图 2.4

2.1.2 Selenium IDE 的功能界面与工具栏

单击界面上的"Create a new project",创建一个新的测试项目,进入 Selenium IDE 的工作界面,其主要区域介绍如图 2.5 所示。

第 2 章　Selenium IDE 的使用

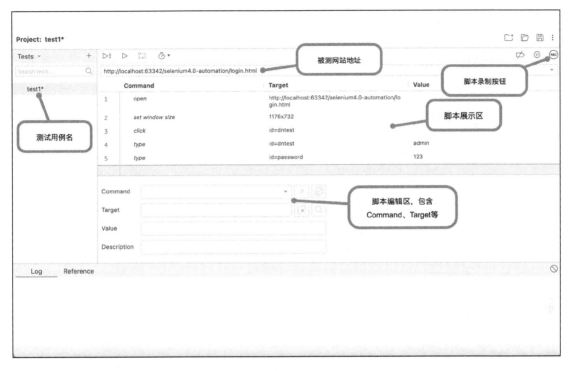

图 2.5

工具栏工具介绍如下。

- ▷≡：执行全部测试用例。

- ▷：执行当前测试用例。

- 8⌐]：单步执行。

- ⏱▼：测试用例执行速度的控制器，可以从 Fast 到 Slow 进行设置。

- ⌀：去除所有断点。

- ⏸：暂停测试用例的执行。

- REC：录制测试用例。

2.1.3　Selenium IDE 脚本

本节介绍录制脚本和增强脚本的方法。本节利用自建 HTML 页面来演示脚本录制。创建名为 login.html 的文件，内容如下：

```html
<!DOCTYPE html>
<html lang="en">
<head>
    <meta charset="UTF-8">
    <title>大牛测试</title>
    <script src="https://***.staticfile.org/jquery/1.10.2/jquery.min.js"></script>
</head>
<h1>大牛自动化测试</h1>
<body>
<form name="daniu">
    <label for="username">用户名：</label>
    <input type="text" name="daniu" id="dntest" class="f-text phone-input" value="大牛" size="22">
    <label for="password">密码：</label>
    <input type="password" name="password" id="password" class ="passwd">
    <button type="submit" id ="loginbtn">登录</button>
    <br />
    <br>
    <a href="upload.html" target="_blank">上传资料页面</a>
    <br>
    <br>
    <label >
      <input type="radio" name="checkbox1"  value="gril" checked>女
    </label>
    <label>
      <input type="radio" name="checkbox2"  value="boy">男
    </label>
</form>
</body>
</html>
```

录制脚本的步骤如下。

（1）单击录制按钮，首先需要设置项目 URL，如"http://localhost:63342/selenium4.0-automation/login.html"，具体如图 2.6 所示。

图 2.6

（2）单击录制按钮，开始录制。

（3）在登录页面的"用户名"输入框中输入"admin"，"密码"输入框中输入"123"。录制结束，录制界面如图 2.7 所示。

图 2.7

（4）脚本展示区中有 5 行数据，分别是：第 1 行显示 Command 值为"open"，Target 值为"http://localhost:63342/selenium4.0-automation/login.html"；第 2 行显示 Command 值为"set window size"，Target 值为"1176x732"；第 3 行显示 Command 值为"click"，Target 值为"id=dntest"；第 4 行显示 Command 值为"type"，Target 值为"id=dntest"，Value 值为"admin"；第 5 行显示 Command 值为"type"，Target 值为"id=password"，Value 值为"123"。

（5）保存脚本，名称为"se_ide1.side"，单击执行当前测试用例按钮，执行脚本，执行成功，在 Log 区域中会显示"'×××' completed successfully"字样，具体如图 2.8 所示。

图 2.8

增强脚本，以验证登录页面上是否存在"登录"按钮为例，步骤如下。

（1）验证登录页面上的元素：按钮"登录"。

（2）添加 assertText 要素到之前录制的脚本中，如图 2.9 所示，添加的 Command 是"assert text"，Target 是"id=loginbtn"，Value 是"登录"。这说明此时的检查点是检查页面上是否显示"登录"字符串。如果有，则检查通过，脚本继续执行；如果没有，则检查未通过，脚本停止执行。如图 2.9 所示的 Log 区域显示，脚本执行和检查点检查都成功了。

第 2 章　Selenium IDE 的使用

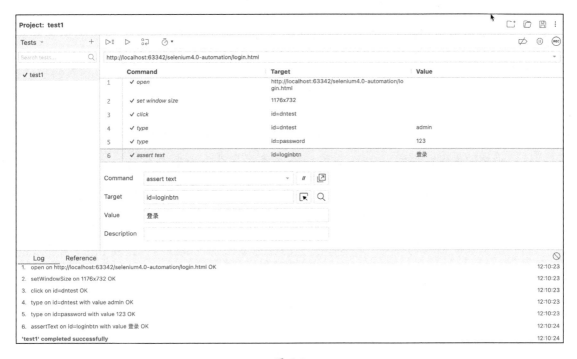

图 2.9

2.1.4　wait for text、assert text 和 verify text 命令

下面分别介绍一下 wait for text、assert text 和 verify text 命令。

- wait for text：从字面意思上理解，该命令用来判断指定文本是否在页面上显示。如果显示，则脚本继续执行；如果在等待一段时间后，没有显示指定文本，则标记脚本执行失败，但测试脚本会继续执行。

- assert text：该命令表示在执行测试脚本时，判断页面上的文本是否与期望显示的文本相同。如果相同，则测试脚本继续执行；如果不同，则标记脚本执行失败，且脚本后续部分不会继续执行。

- verify text：该命令表示在执行测试脚本时，判断页面上的文本是否与期望显示的文本相同。如果相同，则测试脚本继续执行；如果不同，则标记脚本执行失败，而脚本后续部分会继续执行。

2.1.5 通过实例讲解 store title 和 echo 命令

store title 和 echo 命令的作用介绍如下。

- store title 命令的作用是，将网页的 title 属性值存储到指定的变量中。
- echo 命令的作用是，在控制台打印输出内容，常用于脚本调试过程。

以登录页面为例，讲解 store title 和 echo 命令的用法，步骤如下。

（1）打开 Selenium IDE。

（2）在地址栏中输入登录页面地址。

（3）增强优化脚本展示区，具体如图 2.10 所示，添加"open"、"store title"和"echo"命令。

	Command	Target	Value
1	✓ open	http:/localhost:63342/selenium4.0-automation/login.html	
2	✓ store title		title1
3	✓ echo	${title1}	

图 2.10

（4）执行脚本，Log 区域如图 2.11 所示，浏览器窗口的 title 属性打印成功。

```
Log    Reference
Running 'test2'                                                          12:24:50
1.  open on http:/localhost:63342/selenium4.0-automation/login.html OK    12:24:50
2.  storeTitle with value title1 OK                                       12:24:50
echo: 大牛测试                                                             12:24:51
'test2' completed successfully                                            12:24:51
```

图 2.11

2.2 从 Selenium IDE 导出脚本

Selenium IDE 工具的一个重要功能是，可以导出录制过程，生成多种编程语言脚本。

通过录制脚本到自动化脚本的转换，可以提高工程师的脚本编写效率。下面通过 2.1.5 节中的例子来演示从 Selenium IDE 导出脚本并将其运用在自动化测试中的过程，步骤如下。

（1）在你要导出的测试用例上单击鼠标右键，在弹出的快捷菜单中选择"Export"，具体如图 2.12 所示。

图 2.12

（2）然后选中"Python pytest"单选项，具体如图 2.13 所示。

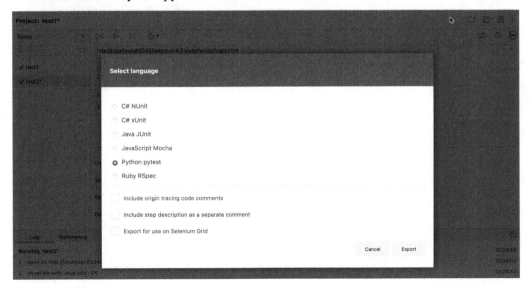

图 2.13

（3）保存.py 文件，自动化脚本如下：

```python
import pytest
import time
import json
from selenium import webdriver
from selenium.webdriver.common.by import By
from selenium.webdriver.common.action_chains import ActionChains
from selenium.webdriver.support import expected_conditions
from selenium.webdriver.support.wait import WebDriverWait
from selenium.webdriver.common.keys import Keys
from selenium.webdriver.common.desired_capabilities import DesiredCapabilities
class TestTest2():
  def setup_method(self, method):
    self.driver = webdriver.Firefox()
    self.vars = {}
  def teardown_method(self, method):
    self.driver.quit()
  def test_test2(self):
    self.driver.get("http://localhost:63342/selenium4.0-automation/login.html")
    self.vars["title1"] = self.driver.title
    print("{}".format(self.vars["title1"]))
```

通过以上实例的演示，可以看到 Selenium IDE 适合在一些比较简单的自动化测试项目中使用。但对于相对长期的、稳定的和复杂的自动化项目来讲，不太适合使用 Selenium IDE 录制脚本的方式。

第 3 章 Python 与 Selenium 环境搭建

　　Python 是一门跨平台的、开源免费的、解释型的、面向对象的编程语言，近些年比较热门，在 AI（人工智能）、机器学习方面的应用特别广泛。同样，在自动化测试领域，在 Selenium 开发语言的选择上，Python 也很受青睐，这门语言的优势如下。

- 结构简单、语法清晰、关键字少，初学者可以在短时间内上手。
- 有广泛的标准库支持。这也是 Python 的优势之一，丰富的代码库可以让开发者更专注于业务。
- 代码的可扩展性强。比如，对于基于 Python 的应用，如果不想将一些算法等代码结构公开，可以利用 C/C++编写这些代码结构，然后从 Python 代码中调用，编码过程和选择都很灵活。
- 数据库接口丰富。Python 为 MySQL、SQLite 和 MongoDB 等数据库提供了很好的接口支持。

- 可移植性好。

本书选择 Python 3.x 进行讲解，主要基于以下原因。

（1）编码方式和运行效率较之前的版本有所提升。

（2）Python 2 已在 2020 年停用。

在开启我们的自动化测试之旅前，首先要搭建基础的软件环境。本章将主要介绍两种环境的搭建，一种是 Windows 环境，另一种是 macOS 环境。

3.1　Windows 环境下的安装

3.1.1　安装 Python

安装 Python 的步骤如下。

（1）访问 Python 官网，下载相应的软件包。本书中的案例使用的是 64 位系统，所以选择"Download Windows installer (64-bit)"下载，如图 3.1 所示。

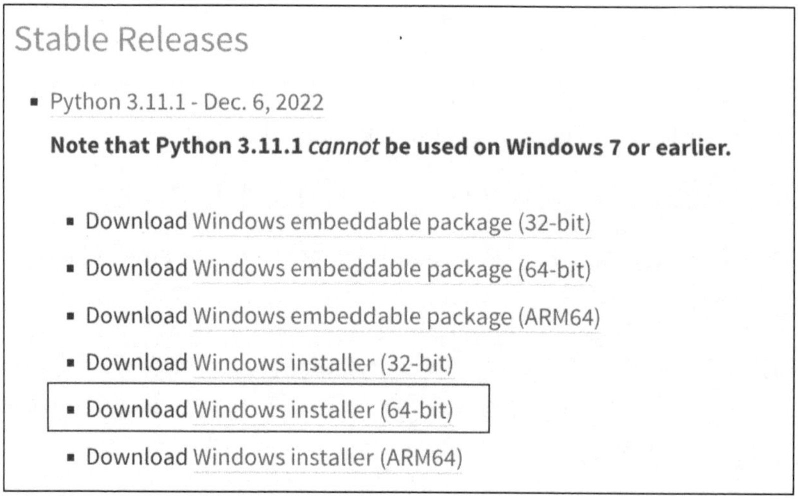

```
Stable Releases

• Python 3.11.1 - Dec. 6, 2022
  Note that Python 3.11.1 cannot be used on Windows 7 or earlier.

  ▪ Download Windows embeddable package (32-bit)
  ▪ Download Windows embeddable package (64-bit)
  ▪ Download Windows embeddable package (ARM64)
  ▪ Download Windows installer (32-bit)
  ▪ Download Windows installer (64-bit)
  ▪ Download Windows installer (ARM64)
```

图 3.1

（2）环境变量。将 Python 3.11 添加到环境变量 PATH 中，这样安装完毕后无须手动配置环境变量，如图 3.2 所示。如果选择"Install Now"，则会立即安装。

图 3.2

（3）如果第二步选择"Customize installation"进行安装，则会显示安装预选信息，按照默认的方式设置即可，如图 3.3 所示。直接在界面上单击"Next"按钮，即可进行安装。

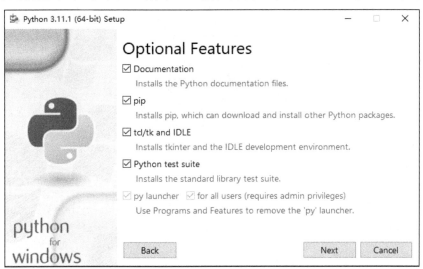

图 3.3

（4）验证 Python 安装得是否正确。

在 DOS 窗口中，输入"python"，在命令行窗口中会显示当前系统安装的默认 Python 版本信息等，如图 3.4 所示。

```
C:\Users\lijin>python
Python 3.11.1 (tags/v3.11.1:a7a450f, Dec  6 2022, 19:58:39) [MSC v.1934 64 bit (AMD64)] on win32
Type "help", "copyright", "credits" or "license" for more information.
>>>
```

图 3.4

图 3.4 中的窗口是 Python 解释器的窗口，可以直接执行代码，比如输入"2+3"，然后按下回车键，窗口中会显示"5"（2+3 的计算结果），具体如图 3.5 所示。

```
C:\Users\lijin>python
Python 3.11.1 (tags/v3.11.1:a7a450f, Dec  6 2022, 19:58:39) [MSC v.1934 64 bit (AMD64)] on win32
Type "help", "copyright", "credits" or "license" for more information.
>>> 2+3
5
>>>
```

图 3.5

Python 安装完成后，用户需要简单了解 Python 自带的一些工具。IDLE 是 Python 自带的编辑器，也是其默认的编程环境。IDLE 是一个 Python Shell，用户可以通过它执行 Python 命令。如果命令执行顺利，会立即看到执行结果，否则会抛出异常。IDLE 位于"开始→程序→Python 3.11"选项中，如图 3.6 所示。

图 3.6

IDLE 的使用步骤如下。

（1）打开 IDLE，选择"File→New File"，输入代码"print("python")"，然后将文件保存为.py 或者.pyw 格式。

（2）单击菜单栏上的"Run→Run Model F5"。

（3）IDLE Shell 窗口中将会显示文本"python"。

具体如图 3.7 所示。

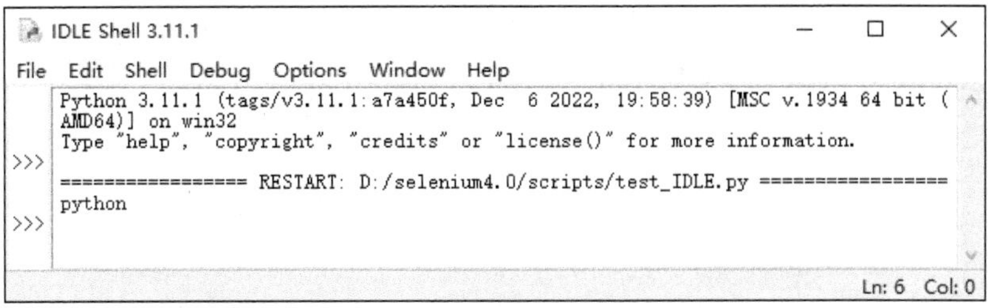

图 3.7

3.1.2 安装 Selenium

Selenium 的安装有两种方式。

1. 通过 pip 工具安装

pip 是安装和管理 Python 包的工具，新版本 Python 中已集成 pip 库。在 DOS 窗口中，键入"pip"，会显示如图 3.8 所示的提示信息。

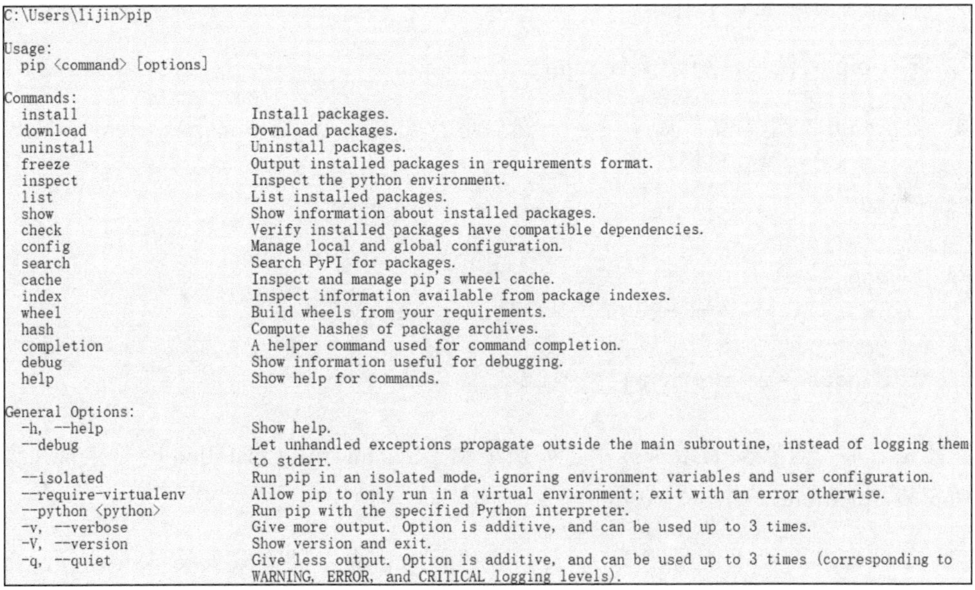

图 3.8

安装 Selenium 之前应先查看本机是否安装了 Selenium，在 DOS 窗口中输入"pip show selenium"，按下回车键后发现出现警告信息"Package(s) not found: selenium"，说明本机还没有安装 Selenium，如图 3.9 所示。

```
C:\Users\lijin>pip show selenium
WARNING: Package(s) not found: selenium

C:\Users\lijin>
```

图 3.9

使用 pip 安装 Selenium 的命令是"pip install selenium"，输入该命令后系统就可以自动安装 Selenium 了。默认情况下，会安装最新版 Selenium。

选择国内网络进行安装时，网络速度一般比较慢，这时可以通过改变 pip 源的方式进行安装，一般有两种方式：永久方式和临时方式。

（1）永久方式。具体设置方式如下（以阿里云提供的镜像源为例）。

① 确认 Windows 目录"C:\Users\[用户名]\AppData\Roaming"下有没有 pip 文件夹，如果没有，则需要新建。

② 进入 pip 文件夹，新建文件 pip.ini。

③ 在 pip.ini 文件中加入如下内容，然后运行安装命令"pip install selenium"：

```
[global]
#超时设定
timeout = 10000
#指定下载源
index-url = http://mirrors.******.com/pypi/simple/
#指定域名
trusted-host = mirrors.aliyun.com
```

安装完之后，查看安装结果提示，如果出现"Successfully installed ××"字样，说明已经成功安装 Selenium，如图 3.10 所示。

（2）临时方式。可以使用在 DOS 窗口中执行 pip 命令的方式安装 Selenium，命令为"pip -i ××"，其中××是镜像源。

```
Collecting selenium==4.7.2
  Using cached selenium-4.7.2-py3-none-any.whl (6.3 MB)
Requirement already satisfied: urllib3[socks]~=1.26 in d:\software\311\lib\site-packages (from selenium==4.7.2) (1.26.14
)
Requirement already satisfied: trio~=0.17 in d:\software\311\lib\site-packages (from selenium==4.7.2) (0.22.0)
Requirement already satisfied: trio-websocket~=0.9 in d:\software\311\lib\site-packages (from selenium==4.7.2) (0.9.2)
Requirement already satisfied: certifi>=2021.10.8 in d:\software\311\lib\site-packages (from selenium==4.7.2) (2022.12.7
)
Requirement already satisfied: attrs>=19.2.0 in d:\software\311\lib\site-packages (from trio~=0.17->selenium==4.7.2) (22
.2.0)
Requirement already satisfied: sortedcontainers in d:\software\311\lib\site-packages (from trio~=0.17->selenium==4.7.2)
(2.4.0)
Requirement already satisfied: async-generator>=1.9 in d:\software\311\lib\site-packages (from trio~=0.17->selenium==4.7
.2) (1.10)
Requirement already satisfied: idna in d:\software\311\lib\site-packages (from trio~=0.17->selenium==4.7.2) (3.4)
Requirement already satisfied: outcome in d:\software\311\lib\site-packages (from trio~=0.17->selenium==4.7.2) (1.2.0)
Requirement already satisfied: sniffio in d:\software\311\lib\site-packages (from trio~=0.17->selenium==4.7.2) (1.3.0)
Requirement already satisfied: cffi>=1.14 in d:\software\311\lib\site-packages (from trio~=0.17->selenium==4.7.2) (1.15.
1)
Requirement already satisfied: wsproto>=0.14 in d:\software\311\lib\site-packages (from trio-websocket~=0.9->selenium==4
.7.2) (1.2.0)
Requirement already satisfied: PySocks!=1.5.7,<2.0,>=1.5.6 in d:\software\311\lib\site-packages (from urllib3[socks]~=1.
26->selenium==4.7.2) (1.7.1)
Requirement already satisfied: pycparser in d:\software\311\lib\site-packages (from cffi>=1.14->trio~=0.17->selenium==4.
7.2) (2.21)
Requirement already satisfied: h11<1,>=0.9.0 in d:\software\311\lib\site-packages (from wsproto>=0.14->trio-websocket~=0
.9->selenium==4.7.2) (0.14.0)
Installing collected packages: selenium
Successfully installed selenium-4.7.2
```

图 3.10

用 pip 安装指定版本的 Selenium 的命令是"pip install selenium==4.7.2",即安装版本号为 4.7.2 的 Selenium。

用 pip 升级最新版的 Selenium 的命令是"pip install --upgrade selenium",即将当前的 Selenium 版本升级到最新版,当前版本会被覆盖。

用 pip 卸载 Selenium 的命令是"pip uninstall selenium"。

pip 工具常用命令总结如表 3.1 所示。

表 3.1

pip 命令	命 令 解 释
pip download 软件包名[==版本号]	下载扩展库的指定版本,如果未指定版本号,则下载扩展库中的最新版
pip list	列出当前环境下所有已经安装的模块
pip install 软件包名[==版本号]	在线安装指定版本的软件包,如果未指定版本号,则下载最新版
pip install 软件包名.whl	通过 whl 离线安装文件进行安装
pip install 包1 包2 包3 …	支持在线依次安装包1、包2等,包名之间用空格隔开
pip install -r list.txt	依次安装在 list.txt 中指定的扩展包
pip install --upgrade 软件包	升级软件包
pip uninstall 软件包名[==版本号]	卸载指定版本的软件包

2. 通过官方离线包安装

访问 Selenium 官方网站，直接下载 Selenium 官方离线包并进行解压，目录结构如图 3.11 所示。

图 3.11

在 DOS 窗口中切换到 Selenium 安装包主目录，执行命令"python setup.py install"即可进行自动安装。

3.1.3 安装开发工具和 IDE

通过上面烦琐的配置后，我们终于搭建好了自动化测试环境，你一定迫不及待地要跟着笔者一起写自动化脚本了吧。别急！在此之前，我们需要找到合适的 IDE（Integrated Developent Environment，集成开发环境）。

1. 安装 Visual Studio Code

Visual Studio Code（以下简称 VS Code）是一款轻量级但功能很强大的源码编辑器软件，能够打开和编辑多种代码文件，可以轻松支持字符编码，方便管理多种多样的插件，应用十分广泛。

Windows 版本 VS Code 软件的安装比较简便，运行安装文件后，在安装界面上直接单击"下一步"按钮，按照提示进行安装即可，如图 3.12 所示。

安装完成后，在本地新建一个文件夹，如"python_test"，然后打开 VS Code，打开新建的文件夹，在该文件夹下新建 Python 文件 test1.py。具体如图 3.13 所示，执行结果如预期一样，可以输出字符"python"。

第 3 章　Python 与 Selenium 环境搭建

图 3.12

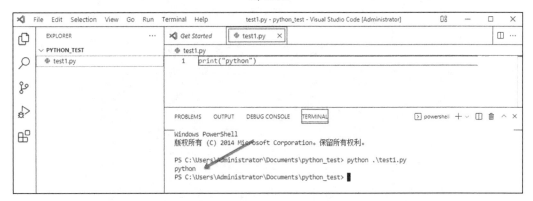

图 3.13

2．安装 PyCharm

PyCharm 是一款 Python 开发中应用比较广泛的 IDE 软件，界面友好，功能强大，还可以跨平台。PyCharm 的优点包括完美支持对 MetaClass 的分析、Refactor、类型猜测、代码补全、代码调试等。可以在 PyCharm 官网上选择 PyCharm 社区版（免费）进行下载并安装，如图 3.14 所示。

运行 PyCharm 安装文件后，在安装界面上单击"Next"按钮，按照提示进行安装即可。

图 3.14

安装完成后，创建新项目，如图 3.15 所示。

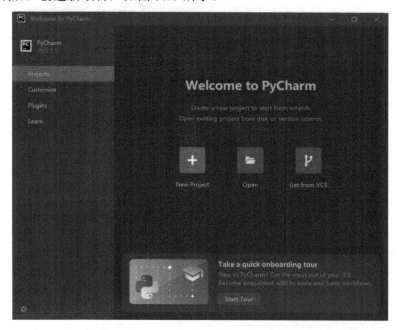

图 3.15

设置项目路径和项目工程文件名，创建项目。然后选中项目工程文件名，单击鼠标右键后在弹出的快捷菜单中选择"New→Python File"，便可开始代码之旅，如图 3.16 所示。

第 3 章　Python 与 Selenium 环境搭建

图 3.16

在第一次使用 PyCharm 时，需要对其进行基本的设置，这可以在一定程度上提高编码效率，设置步骤如下。

（1）设置 PyCharm 的默认解释器。依次选择 "File→Settings…"，打开 Settings 界面，如图 3.17 所示。可以根据不同的项目选择不同的解释器。如果本地安装了多个版本的 Python，那么针对不同的项目也可以选择不同版本的 Python 解释器。

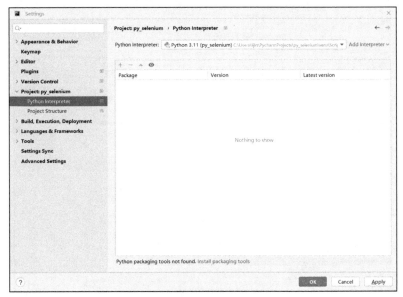

图 3.17

（2）设置 PyCharm 的默认文件编码格式，如图 3.18 所示，全局文件编码格式为"UTF-8"。

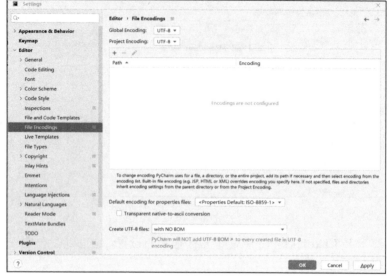

图 3.18

（3）设置 PyCharm 的代码格式。选择"Code Style"，设置自己需要的代码格式，如 Python、HTML、JSON 和 XML 等，设置如图 3.19 所示。

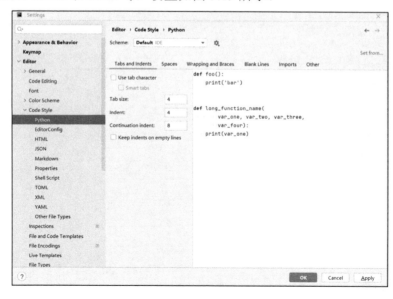

图 3.19

(4)设置默认的 XML 模式,如图 3.20 所示,当前 HTML 的默认设置为 HTML 5 语法。

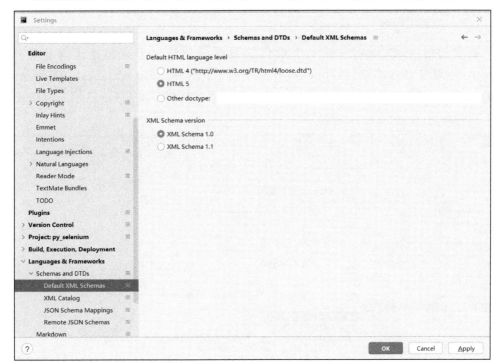

图 3.20

PyCharm 常用的快捷键如下(其中符号"+"不是加号键,而是表示"和"的意思)。

- 注释代码:选中代码后,按快捷键 Ctrl + /(在 macOS 环境下,按快捷键 command + /)。
- 定位声明等信息:将鼠标光标放在代码处,按快捷键 Ctrl + 鼠标左键(在 macOS 环境下,按快捷键 command + 鼠标左键)。
- 代码缩进:Tab 键。
- 复制选定的区域或行:快捷键 Ctrl + D(在 macOS 环境下,按快捷键 command + D)。
- 删除选定的行:快捷键 Ctrl + Y(在 macOS 环境下,按快捷键 command + Y)。
- Shift + F10:运行代码。
- Shift + F9:调试代码。

- F8：调试代码——跳过。
- F7：调试代码——进入。

以上我们对 PyCharm 进行了个性化的设置，如果出于某种原因需要重新安装 PyCharm，可以将 IDE 的设置导出。如图 3.21 所示，选择"File→Manage IDE Settings→Export Settings…"。可以在导出设置时选择不同的导出格式，如以 zip 压缩文件格式导出，如图 3.22 所示。

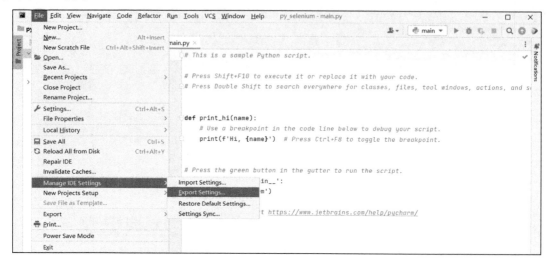

图 3.21

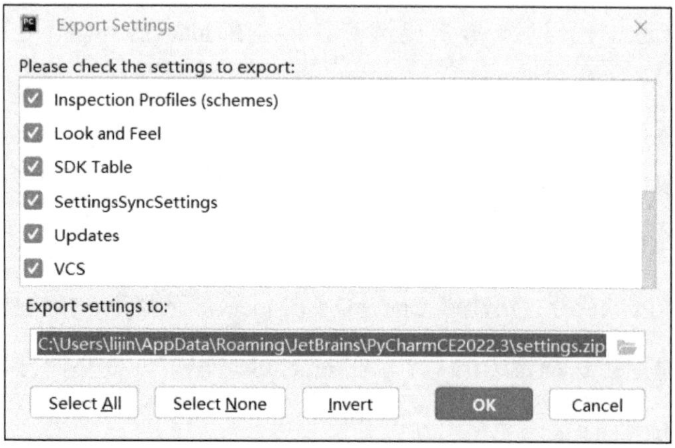

图 3.22

3.1.4 搭建不同的浏览器环境

1. Edge

搭建 Edge 浏览器环境，需要下载对应的浏览器驱动。Selenium 的浏览器驱动都可以通过 Selenium 官网下载，如图 3.23 所示。因国外网站下载速度慢，可以选择国内的 taobao 镜像源。

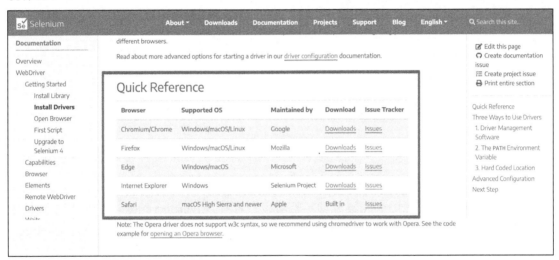

图 3.23

下载完成后，将对应的浏览器驱动（驱动程序名 msedgedriver.exe）放到环境变量中或者指定的路径下。在 PyCharm 中新建文件 testlogin.py，并添加如下代码，这是一种显式指定驱动路径的方式，实现了简单的打开 Edge 浏览器并访问 login.html 页面的操作。

```python
#引入Python自带的os代码模块
import os
#导入WebDriver模块和Service类
#其中Service类是Selenium 4版本引入的新用法
from selenium import webdriver
from selenium.webdriver.ie.service import Service
#设置Edge浏览器的WebDriver驱动程序路径
edge_driver_server = Service("C:\\msedgedriver.exe")
```

```
#启动 Edge 浏览器
driver = webdriver.Edge(service=edge_driver_server)
#打开百度首页
driver.get('file:///Users/tim/Desktop/selenium/book/login.html')
```

2. IE

搭建 IE 浏览器环境,需要下载对应的 IE 浏览器驱动。Selenium 的浏览器驱动都可以通过 Selenium 官网下载,如图 3.24 所示。

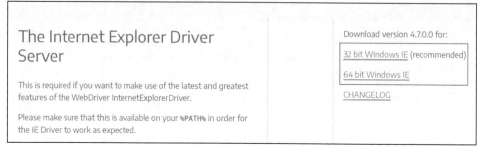

图 3.24

下载完成后,将对应的浏览器驱动(驱动程序名 IEDriverServer.exe)放到环境变量中或者指定的路径下。比如,把 IE 浏览器驱动放在 "C:\Program Files (x86)\Internet Explorer" 目录下,在脚本中需要指定路径;也可以直接将浏览器驱动放到系统环境变量中,这时不需要在脚本中指定路径。下面以打开 login.html 页面为例进行环境测试,代码如下:

```
#引入 Python 自带的 os 代码模块
import os
#导入 WebDriver 模块
from selenium import webdriver
#设置 IE 浏览器的 WebDriver 驱动程序路径
IEDriverServer="C:\\Program Files (x86)\\Internet Explorer\\IEDriver Server.exe"
#设置当前 OS WebDriver 为 IE 浏览器的驱动程序
os.environ["webdriver.ie.driver"] = IEDriverServer
#启动 IE 浏览器
driver = webdriver.Ie(IEDriverServer)
#打开 login.html 页面
driver.get('file:///Users/tim/Desktop/selenium/book/login.html')
```

3. Chrome

Chrome 最近几年发展迅猛,市场占有率稳步上升。搭建 Chrome 浏览器环境需要下载与 Chrome 浏览器相匹配的驱动程序 chromedriver.exe。下面以打开 login.html 页面为例,代码如下:

```
from selenium import webdriver
from selenium.webdriver.chrome.service import Service
#这里需要指定真实的Chrome浏览器驱动程序路径
ChromeDriverServer = Service("C:\\Users\\xxx\\chromedriver.exe")
driver = webdriver.Chrome(ChromeDriverServer)
driver.get('file:///Users/tim/Desktop/selenium/book/login.html')
```

4. Firefox

搭建 Firefox 浏览器环境,如果 Firefox 的版本高于 47.0.1,则需要下载浏览器驱动,驱动程序名 geckodriver.exe,其使用方式同 IE 和 Chrome。下面以打开 login.html 页面为例,代码如下:

```
from selenium import webdriver
from selenium.webdriver.firefox.service import Service
path=Service("D:\\dr\\geckodriver.exe")
driver = webdriver.Firefox(executable_path=path)
driver.get("file:///C:/Users/Administrator/Desktop/login.html")
```

3.2 macOS 环境下的安装

关于 macOS 环境下的安装,下面只介绍安装 Python 和 Selenium,其他软件的安装步骤与 Windows 环境下的安装步骤类似,故不再赘述。

3.2.1 安装 Python

macOS 环境下默认已安装了 Python 2,如图 3.25 所示。

```
Python 2.7.10 (v2.7.10:15c95b7d81dc, May 23 2015, 09:33:12)
[GCC 4.2.1 (Apple Inc. build 5666) (dot 3)] on darwin
Type "help", "copyright", "credits" or "license" for more information.
>>>
```

图 3.25

但是我们需要使用 Python 的最新版本,所以下载 macOS 环境下的最新版 Python 安装包,如图 3.26 所示。

图 3.26

下载 Python 安装包后,双击文件"python-3.11.1-macos11.pkg",如图 3.27 所示,单击"继续"按钮,按提示进行安装即可。安装完成后,Python 3 默认路径为 /Library/Frameworks/Python.framework/Versions/3.11。

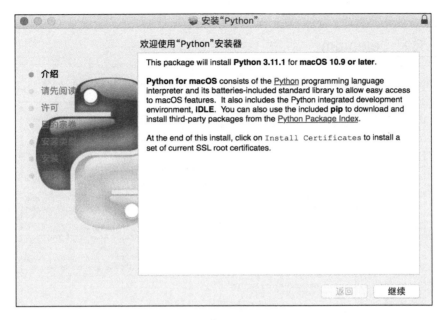

图 3.27

设置 Python 3 为默认版本，先打开终端，输入如下命令打开配置文件：

#vim ~/.bash_profile

在配置文件中加入以下内容：

alias python="/Library/Frameworks/Python.framework/Versions/3.11/bin/python3"

alias pip="/Library/Frameworks/Python.framework/Versions/3.11/bin/pip3"

保存并退出，然后执行以下命令：

#source ~/.bash_profile

在命令行窗口中输入"python"，显示"Python 3.11.1"，说明版本切换成功，如图 3.28 所示。

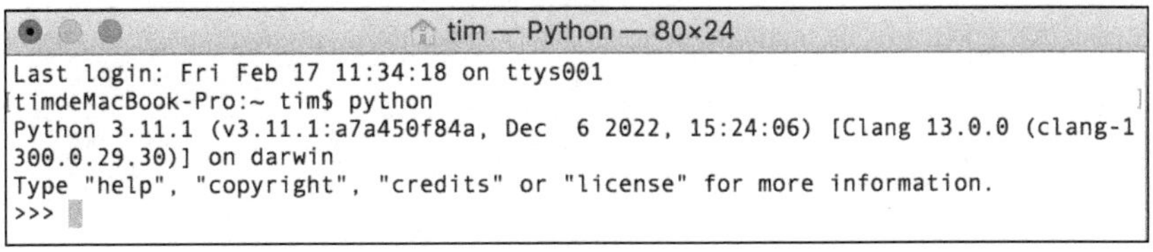

图 3.28

3.2.2 安装 Selenium

使用 pip 安装 Selenium 的命令为 "pip install selenium"，安装界面如图 3.29 所示。

```
m==4.7.2) (1.10)
Requirement already satisfied: idna in /Library/Frameworks/Python.framework/Vers
ions/3.11/lib/python3.11/site-packages (from trio~=0.17->selenium==4.7.2) (3.4)
Requirement already satisfied: outcome in /Library/Frameworks/Python.framework/V
ersions/3.11/lib/python3.11/site-packages (from trio~=0.17->selenium==4.7.2) (1.
2.0)
Requirement already satisfied: sniffio in /Library/Frameworks/Python.framework/V
ersions/3.11/lib/python3.11/site-packages (from trio~=0.17->selenium==4.7.2) (1.
3.0)
Requirement already satisfied: wsproto>=0.14 in /Library/Frameworks/Python.frame
work/Versions/3.11/lib/python3.11/site-packages (from trio-websocket~=0.9->selen
ium==4.7.2) (1.2.0)
Requirement already satisfied: PySocks!=1.5.7,<2.0,>=1.5.6 in /Library/Framework
s/Python.framework/Versions/3.11/lib/python3.11/site-packages (from urllib3[sock
s]~=1.26->selenium==4.7.2) (1.7.1)
Requirement already satisfied: h11<1,>=0.9.0 in /Library/Frameworks/Python.frame
work/Versions/3.11/lib/python3.11/site-packages (from wsproto>=0.14->trio-websoc
ket~=0.9->selenium==4.7.2) (0.14.0)
Installing collected packages: selenium
Successfully installed selenium-4.7.2
```

图 3.29

3.2.3 浏览器的驱动

在 macOS 系统中，驱动一般存放于/usr/bin 目录下。下载对应 macOS 版本的驱动文件后，将其转移到 bin 目录下，便可驱动相应的浏览器，也可以将驱动文件复制到项目工程文件中，如图 3.30 所示。

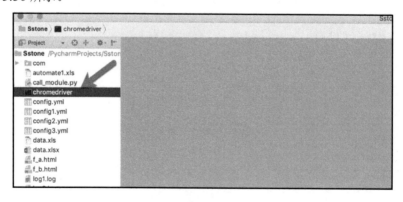

图 3.30

其他软件的安装、使用等与在 Windows 系统中类似，故不再赘述。

第二篇

基　础　篇

古人云："九层之台，起于累土；千里之行，始于足下。"这句话说明了积累很重要，所以如果我们要掌握并熟练运用 Selenium 的核心基础功能，就要多尝试、多实战，这样才能在各种失败或错误中积累经验。本篇对应的章节如下。

第 4 章　Selenium 元素定位

第 5 章　Selenium 常用方法

第 4 章
Selenium 元素定位

自 Selenium 2 之后，WebDriver 就出现在大众的视野中。它利用浏览器原生 API 封装了一些底层操作的功能，使得它作为一套框架更容易被使用。Selenium 支持多种编程语言，如 Python、Java、PHP 等。本书采用 Python 3，在开始自动化测试之前，有必要先了解一些 Python 基础知识，然后学习 Selenium 八大定位、表格定位和关联元素定位策略等知识。

4.1 Python 基础知识

Python 是跨平台、开源免费的高级编程语言之一，在自动化测试行业内应用比较广泛。Python 支持伪编译，可以将 Python 源程序转换为字节码来提高性能。Python 也支持将多语言程序无缝拼装（比如在 Python 代码中调用 C 程序代码），这样可以更好地发挥不同工具或语言的优势来满足不同的需求。本节将简单介绍 Selenium 中经常用到的 Python 基础知识。

在 Python 中，变量不需要事先声明变量名和数据类型，可以直接赋值使用，并且适用于任意类型的对象。变量的类型也可以随时变化。如下面的例子：

```
#x 变量的类型是 int 型，并且值为"3"
x = 3
#变量类型已经变为 str 型
x = "hello"
```

在定义变量名时，要注意的事项如下。

- 变量名应以字母或下画线开头。
- 变量名中不能包含空格及标点符号（括号、引号、逗号、问号、句号等）。
- 不能使用关键字作为变量名。对于关键字，可以在导入 keyword 模块后，执行 print(keyword.kwlist)命令查看。
- 变量名对英文大小写敏感。

在 Python 中，对象模型是一个非常重要的概念，Python 中的一切都可以是对象，比如内置对象和非内置对象。内置对象可以直接调用，非内置对象需要在导入相关的模块后才可以使用。

4.1.1 数字类型

数字类型名称有 int、float、complex，分别代表整型、浮点型和复杂型数字类型。在 Python 中，数字大小没有限制，并且支持复数及其相关计算。如整型 1234、复数型 4+5j。常用的整型数字可以按照进制分为二进制数、八进制数、十进制数和十六进制数。

- 二进制数，必须以"0b"开头，如 0b101 等。
- 八进制数，必须以"0o"开头，如 0o23、0o12 等。
- 十进制数，最常用，如 123、-5、0 等。
- 十六进制数，必须以"0x"开头，如 0x11、0xfb 等。

4.1.2 字符串类型

在 Python 中，字符串可以用双引号或单引号来指定，如"上海"或者'上海'。测试代码如下：

```
#coding=utf-8
from_station = "上海"
print(from_station)
print(from_station[1:2])
print(from_station[0:1])
```

以上代码执行之后的结果如图 4.1 所示，代码的第 3 行和第 4 行执行的是对字符串的切片操作，其作用是分别打印字符串中的第 2 个汉字和第 1 个汉字。

```
/usr/local/bin/python3 /Users/tim/Documents/pyse2023/chapter04/daniu_4.1.2.py
上海
海
上

Process finished with exit code 0
```

图 4.1

在 Python 中，会经常用到特殊字符，和其他开发语言一样，这些特殊字符需要转义。Python 常用的转义字符及含义如图 4.2 所示。

转义字符符号	字符含义
\'	单引号
\"	双引号
\a	发出系统响铃声
\b	退格符
\n	换行符
\t	横向制表符
\v	纵向制表符
\r	回车符
\f	换页符
\o	八进制数转义
\x	十六进制数转义
\000	\000后的字符串全部忽略

图 4.2

4.1.3 常用的判断与循环语句

Python 作为一门编程语言，基本的判断与循环功能是必须的。比较常用的判断语句有"if"。下面以一个简单的例子来说明 Python 中判断语句的用法，代码如下：

```
#coding=utf=8
#首先给一个变量赋值
a = 10
```

```
if a > 10:
    print("数字大于10")
elif a < 10:
    print("数字小于10")
else:
    print("数字等于10")
```

以上代码的输出结果为"数字等于10"。需要注意以下两点。

(1) if、elif 及 else 关键字的写法。

(2) 每个判断语句都是以冒号":"结尾的,注意代码中涉及的标点符号都是英文符号。

以上是 if 判断语句的用法,比较容易掌握。

Python 中的 for 循环语句运用也比较广泛。比如,要处理一批 Python 对象(特别是有关联的对象)时,需要对列表、集合内的所有元素进行处理,这时就需要运用 for 循环语句来提高代码效率。下面举例说明 for 循环语句的用法,代码如下:

```
#coding=utf-8
list1 = ["selenium","appium","python","automation"]
#使用 for 循环语句来遍历列表 list1 中的所有元素
#第一种方式
for l in list1:
    print(l)
#第二种方式
for index in range(len(list1)):
    print(list1[index])
```

以上代码通过两种方式实现了用 for 循环语句遍历列表元素,这两种方式都能实现遍历元素的目的:第一种方式采用直接遍历元素的方式;第二种方式采用通过列表下标的方式来获取列表元素。

其中用到了 range 函数。Python 中的 range 函数功能很强大。在上例中,"len(list1)"的意思是获取列表的长度,如列表的长度为 4;"range(4)"表示由 0、1、2、3 这 4 个数字组成的列表,但其实该函数返回的是一个可迭代对象(range 对象),而不是一个列表,这样也是为了节约内存空间。range 函数在 Python 中运用得非常广泛,语法如下:

```
range(start,stop[,step])
```

参数说明如下。

- start：计数是从 start 开始的，默认是从 0 开始的。
- stop：计数到 stop 结束，但是不包括 stop。
- step：步长，默认为 1。

range 函数的测试代码如下：

```
#coding=utf-8
for i in range(10):
    print(i)
print("***************")
for j in range(2,10,2):
    print(j)
print("***************")
print(type(range(10)))
```

以上测试代码的执行结果如图 4.3 所示，执行结果和预期一致。

```
/usr/local/bin/python3 /Users/tim/Documents/pyse2023/chapter04/daniu_4.1.3.py
0
1
2
3
4
5
6
7
8
9
***************
2
4
6
8
***************
<class 'range'>

Process finished with exit code 0
```

图 4.3

4.1.4 列表对象

在 Python 中，列表是非常重要的概念之一。在运用 Python 进行自动化测试的过程中，列表的应用极其广泛，因此需要重点掌握它的基本用法。列表的类型名称为"list"，与其他编程语言中的数组类似。所有的元素都被一对方括号包含，元素之间使用英文逗号分隔，并且元素可以是不同类型的。Python 列表的功能和 C 语言的数组相似，但是 Python 列表的使用更加灵活简单。在单个列表中，列表元素可以是不同的数据类型，包括整型数字、浮点型数字、字符串和对象等。典型的列表写法如下：

```
list_1 = [3,6,9,"selenium","8.9093",["a","B","C","abc"]]
```

示例：打印列表 list_1。代码如下：

```
#coding=utf-8
list_1 = [3,6,9,"selenium","8.9093",["a","B","C","abc"]]
print("以下为直接打印整个list列表：")
print(list_1)
#换行操作
print("\n")
print("以下为逐个遍历列表的元素，并打印：")
#用for循环语句遍历所有的列表元素，并打印
for l in list_1:
    print(l)
```

以上代码的执行结果如图 4.4 所示。

```
/usr/local/bin/python3 /Users/tim/Documents/pyse2023/chapter04/daniu_4.1.5.py
以下为直接打印整个list列表：
[3, 6, 9, 'selenium', '8.9093', ['a', 'B', 'C', 'abc']]


以下为逐个遍历列表的元素，并打印：
3
6
9
selenium
8.9093
['a', 'B', 'C', 'abc']

Process finished with exit code 0
```

图 4.4

Python 提供了 3 种方法向列表对象中添加元素，分别介绍如下。

（1）append 方法：在列表的最后添加一个元素，并且一次只能添加一个元素。示例代码如下：

```
#coding=utf-8
list_1 = [3,6,9,"selenium","8.9093",["a","B","C","abc"]]
print("append添加列表元素之前,遍历列表元素,并打印")
#用for循环语句遍历所有的列表元素,并打印
for l in list_1: #for
    print(l)
list_1.append("a_append")
#换行操作
print("\n")
print("append添加列表元素之后,遍历列表元素,并打印")
#用for循环语句遍历所有的列表元素,并打印
for l in list_1:
    print(l)
```

以上代码的执行结果如图 4.5 所示。

```
/usr/local/bin/python3 /Users/tim/Documents/pyse2023/chapter04/daniu_4.1.5_1.py
append添加列表元素之前,遍历列表元素,并打印
3
6
9
selenium
8.9093
['a', 'B', 'C', 'abc']

append添加列表元素之后,遍历列表元素,并打印
3
6
9
selenium
8.9093
['a', 'B', 'C', 'abc']
a_append

Process finished with exit code 0
```

图 4.5

(2) extend 方法:一次添加多个元素,新添加的元素在列表最后的位置。示例代码如下:

```
#coding=utf-8
list_1 = [3,6,9,"selenium","8.9093",["a","B","C","abc"]]
print("extend添加列表元素之前,遍历列表元素,并打印")
#用for循环语句遍历所有的列表元素,并打印
```

```
for l in list_1:
    print(l)
list_1.extend(['e','f','g'])
#换行操作
print("\n")
print("extend添加列表元素之后，遍历列表元素，并打印")
#用for循环语句遍历所有的列表元素，并打印
for l in list_1:
    print(l)
```

以上代码的执行结果如图4.6所示。

```
/usr/local/bin/python3 /Users/tim/Documents/pyse2023/chapter04/daniu_4.1.5_2.py
extend添加列表元素之前，遍历列表元素，并打印
3
6
9
selenium
8.9093
['a', 'B', 'C', 'abc']

extend添加列表元素之后，遍历列表元素，并打印
3
6
9
selenium
8.9093
['a', 'B', 'C', 'abc']
e
f
g

Process finished with exit code 0
```

图 4.6

（3）insert方法：在特定位置处添加元素。这里的位置是指元素在列表中的位置索引号。注意，索引号是从0开始的。示例代码如下：

```
#coding=utf-8
list_1 = [3,6,9,"selenium","8.9093",["a","B","C","abc"]]
print("insert添加列表元素之前，遍历列表元素，并打印")
#用for循环语句遍历所有的列表元素，并打印
for l in list_1:
    print(l)
#指在列表的第1个位置（位置索引号为0）处添加元素0
list_1.insert(0,"0")
```

```
#换行操作
print("\n")
print("insert 添加列表元素之后，遍历列表元素，并打印")
#用 for 循环语句遍历所有的列表元素，并打印
for l in list_1:
    print(l)
```

以上代码的执行结果如图 4.7 所示，实现了在列表的第 1 个位置处添加元素"0"。

```
/usr/local/bin/python3 /Users/tim/Documents/pyse2023/chapter04/daniu_4.1.5_3.py
insert添加列表元素之前，遍历列表元素，并打印
3
6
9
selenium
8.9093
['a', 'B', 'C', 'abc']

insert添加列表元素之后，遍历列表元素，并打印
0
3
6
9
selenium
8.9093
['a', 'B', 'C', 'abc']

Process finished with exit code 0
```

图 4.7

删除列表元素的操作在 Selenium 自动化测试中也经常用到，下面介绍常用的 3 种方法。

（1）remove 方法：删除列表中的特定元素。例如，假定在列表 list_1 中有一个元素值是"3"，如果要删除这个元素，代码可以写成"list_1.remove(3)"。示例代码如下：

```
#coding=utf-8
list_1 = [3,6,9,"selenium","8.9093",["a","B","C","abc"]]
print("删除列表元素之前，遍历列表元素，并打印")
#用 for 循环语句遍历所有的列表元素，并打印
for l in list_1:
    print(l)
#删除元素值为 3 的列表元素
list_1.remove(3)
#换行操作
print("\n")
```

```
print("删除列表元素之后,遍历列表元素,并打印")
#用for循环语句遍历所有的列表元素,并打印
for l in list_1:
    print(l)
```

以上代码的执行结果如图 4.8 所示,成功地删除了元素值为 3 的列表元素。

```
/usr/local/bin/python3 /Users/tim/Documents/pyse2023/chapter04/daniu_4.1.5_4.py
删除列表元素之前,遍历列表元素,并打印
3
6
9
selenium
8.9093
['a', 'B', 'C', 'abc']

删除列表元素之后,遍历列表元素,并打印
6
9
selenium
8.9093
['a', 'B', 'C', 'abc']

Process finished with exit code 0
```

图 4.8

(2)del 方法:删除列表中指定位置处的元素。例如,如果要删除列表 list_1 中位置索引号为 1 的元素,代码可以写成"del list_1[1]"。示例代码如下:

```
list_1 = [3,6,9,"selenium","8.9093",["a","B","C","abc"]]
print("删除列表元素之前,遍历列表元素,并打印")
#用for循环语句遍历所有的列表元素,并打印
for l in list_1:
    print(l)
#删除位置索引号为1的列表元素,也就是列表中的第2个元素
del list_1[1]
#换行操作
print("\n")
print("删除列表元素之后,遍历列表元素,并打印")
#用for循环语句遍历所有的列表元素,并打印
for l in list_1:
    print(l)
```

以上代码的执行结果如图 4.9 所示，成功地删除了位置索引号为 1 的列表元素，也就是列表中的第 2 个元素。

```
/usr/local/bin/python3 /Users/tim/Documents/pyse2023/chapter04/daniu_4.1.5_5.py
删除列表元素之前，遍历列表元素，并打印
3
6
9
selenium
8.9093
['a', 'B', 'C', 'abc']

删除列表元素之后，遍历列表元素，并打印
3
9
selenium
8.9093
['a', 'B', 'C', 'abc']

Process finished with exit code 0
```

图 4.9

（3）pop 方法：将列表中的最后一个元素返回，并将其从列表中删除。示例代码如下：

```
list_1 = [3,6,9,"selenium","8.9093","-9"]
print("删除列表元素之前，遍历列表元素，并打印")
#用 for 循环语句遍历所有的列表元素，并打印
for l in list_1:
    print(l)
pop_res = list_1.pop()
print("\n")
print("pop 方法返回的元素："+pop_res)
#换行操作
print("\n")
print("删除列表元素之后，遍历列表元素，并打印")
#用 for 循环语句遍历所有的列表元素，并打印
for l in list_1:
    print(l)
```

以上代码的执行结果如图 4.10 所示，成功地返回了列表的最后一个元素-9，并将其从列表中删除。

```
/usr/local/bin/python3 /Users/tim/Documents/pyse2023/chapter04/daniu_4.1.5_6.py
删除列表元素之前，遍历列表元素，并打印
3
6
9
selenium
8.9093
-9

pop方法返回的元素：-9

删除列表元素之后，遍历列表元素，并打印
3
6
9
selenium
8.9093

Process finished with exit code 0
```

图 4.10

列表分片是指获取列表中的部分元素并将其作为一个新的列表元素。示例代码如下：

```
list_1 = [3,6,9,"selenium","8.9093","-9"]
print("列表分片之前，遍历列表元素，并打印")
#用 for 循环语句遍历所有的列表元素，并打印
for l in list_1:
    print(l)
#换行操作
print("\n")
#返回的字符串是列表中的第 4 个元素
temp = list_1[3]
print(temp)
#连续分片，返回的是一个新的列表 temp，新列表由老列表的第 3、4 个元素组成
temp = list_1[2:4] print(temp)
print("列表分片之后，遍历列表元素，并打印")
#用 for 循环语句遍历所有的列表元素，并打印
for l in list_1:
    print(l)
```

以上代码执行成功，列表分片操作的执行结果如图 4.11 所示。

```
/usr/local/bin/python3 /Users/tim/Documents/pyse2023/chapter04/daniu_4.1.5_7.py
列表分片之前,遍历列表元素,并打印
3
6
9
selenium
8.9093
-9

selenium
[9, 'selenium']
列表分片之后,遍历列表元素,并打印
3
6
9
selenium
8.9093
-9

Process finished with exit code 0
```

图 4.11

接下来,列举 3 种常用的列表操作符。

(1)+:该操作符的作用是对多个列表直接进行拼接。示例代码如下:

```
list_1 = [3,6,9,"selenium","8.9093","-9"]
list_2 = [1,4,7,"python","9.9999","-10"]
print("遍历列表 1 并打印")
#用 for 循环语句遍历所有的列表元素,并打印
for l in list_1:
    print(l)
print("遍历列表 2 并打印")
#用 for 循环语句遍历所有的列表元素,并打印
for l in list_2:
    print(l)
list_3 = list_1 + list_2
print("遍历拼接后的列表并打印")
for l in list_3:
    print(l)
```

以上代码的执行结果如图 4.12 所示,拼接后的列表是列表 1 和列表 2 的组合。

```
/usr/local/bin/python3 /Users/tim/Documents/pyse2023/chapter04/daniu_4.1.5_8.py
遍历列表1并打印
3
6
9
selenium
8.9093
-9
遍历列表2并打印
1
4
7
python
9.9999
-10
遍历拼接后的列表并打印
3
6
9
selenium
8.9093
-9
1
4
7
python
9.9999
-10
```

图 4.12

（2）*：该操作符的作用是实现列表成倍的复制和添加。示例代码如下：

```
list_1 = [3,6,9,"selenium","8.9093","-9"]
print("遍历列表1并打印")
#用for循环语句遍历所有的列表元素，并打印
for l in list_1:
    print(l)
list_3 = list_1*3
print("遍历拼接后的列表并打印")
for l in list_3:
    print(l)
```

以上代码的执行结果如图 4.13 所示，将列表 list_1 中的元素复制 3 倍后返回新列表。

```
/usr/local/bin/python3 /Users/tim/Documents/pyse2023/chapter04/daniu_4.1.5_9.py
遍历列表1并打印
3
6
9
selenium
8.9093
-9
遍历拼接后的列表并打印
3
6
9
selenium
8.9093
-9
3
6
9
selenium
8.9093
-9
3
6
9
selenium
8.9093
-9

Process finished with exit code 0
```

图 4.13

（3）>和<：这两个操作符的作用是比较数据型列表的元素。示例代码如下：

```
list_1 = [1,2,3,4,5]
list_2 = [6,7,8,9,10]
print("遍历列表 1 并打印")
#用 for 循环语句遍历所有的列表元素，并打印
for l in list_1:
    print(l)
print("遍历列表 2 并打印")
#用 for 循环语句遍历所有的列表元素，并打印
for l in list_2:
    print(l)
print(list_1 > list_2)
print(list_1 < list_2)
```

以上代码的执行结果如图 4.14 所示，list_1 的元素小于 list_2 的。

```
/usr/local/bin/python3 /Users/tim/Documents/pyse2023/chapter04/daniu_4.1.5_10.py
遍历列表1并打印
1
2
3
4
5
遍历列表2并打印
6
7
8
9
10
False
True

Process finished with exit code 0
```

图 4.14

4.2 Selenium 八大定位

以上简要介绍了本篇涉及的 Python 基础知识，其他一些基础知识将分散到项目篇中进行讲解。Python 编程所需的技能需要在实践中不断充实和完善。

在 Selenium 中，根据 HTML 页面元素的属性来定位。在 Web 自动化测试过程中，常用的操作步骤如下。

（1）定位网页上的页面元素，并获取元素对象。

（2）对元素对象实施单击、双击、拖曳或输入值等操作。

Selenium 提供了 8 种不同的定位方法，分别通过 id、name、class、link_text、partial_link_text、CSS、XPath 及 tag_name 进行定位。在 Selenium 4 版本中，定位方法 find_element_by_xx 被丢弃，而采用 find_element 方法。具体的使用细节将在本节中详细介绍。

4.2.1 id 定位

在学习 Selenium 元素定位之前，需要对浏览器驱动进行全局设置或管理。一种比较简单方便的方式是，将 Chrome 浏览器驱动程序 chromedriver.exe 放到项目目录中，这样就不需要定义驱动文件路径了（例如 path="C:\Users\lijin\AppData\Local\Google\Chrome\Application\chromedriver.exe"）。以 Chrome 浏览器为例，初始化代码可以被简化为 "driver = webdriver.Chrome()"。本书后面内容中的相关写法，也以这种方式为准。

HTML Tag 的 id 属性值是唯一的，故不存在根据 id 定位多个元素的情况。下面以在 login.html 页面用户名文本框中输入"大牛测试"为例，用户名 id 属性值为"dntest"，输入前如图 4.15 所示。

图 4.15

代码如下，通过 find_element 方法来定位用户名文本框，需要传入两个参数，By.ID 即通过 id 定位元素，另一个参数是 id 值：

```
#coding=utf-8
from selenium import webdriver
from selenium.webdriver.chrome.service import Service
from selenium.webdriver.common.by import By
path =Service('D:\\dr\\chromedriver.exe')
driver = webdriver.Chrome(service=path)
#打开登录页面
driver.get('file:///D:/selenium/book/login.html')
#用户名输入测试
driver.find_element(By.ID,"dntest").send_keys("大牛测试")
```

运行代码后，在用户名文本框中输入了"大牛测试"，如图 4.16 所示。

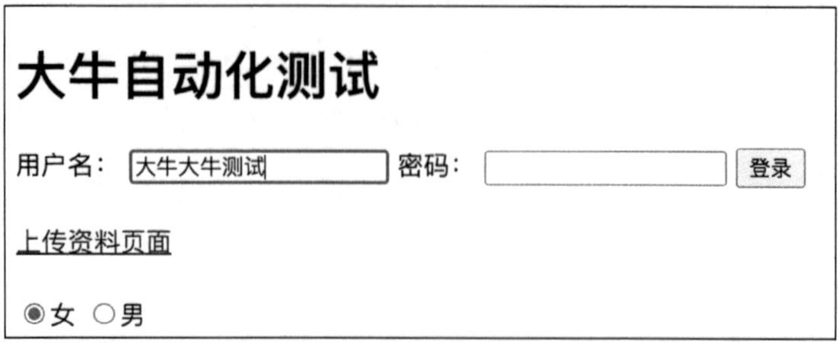

图 4.16

4.2.2 name 定位

以上用户名文本框也可以用 name 定位，其 name 属性值为"daniu"，参数 By.NAME 表示通过 name 定位，代码如下：

```
#coding=utf-8
from selenium import webdriver
from selenium.webdriver.chrome.service import Service
from selenium.webdriver.common.by import By
path =Service('D:\\dr\\chromedriver.exe')
driver = webdriver.Chrome(service=path)
#打开登录页面
driver.get('file:///Users/tim/Desktop/selenium/book/login.html')
#用户名输入测试
driver.find_element(By.NAME,"daniu").send_keys("大牛测试")
```

运行代码后的效果如图 4.17 所示。

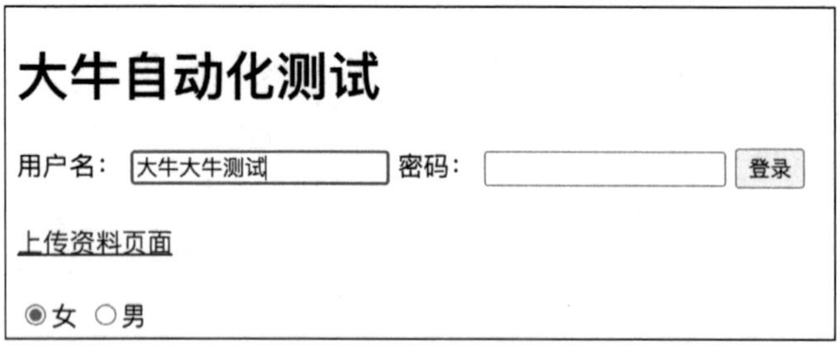

图 4.17

注意：使用 name 定位需要保证 name 属性值唯一，否则定位将失败。

4.2.3 class 定位

以 login.html 密码框为例,其 class 属性值为"passwd",在 find_element 方法中一个参数为 By.CLASS_NAME,另一参数为 passwd,代码如下:

```
#coding=utf-8
from selenium import webdriver
from selenium.webdriver.chrome.service import Service
from selenium.webdriver.common.by import By
path =Service('D:\\dr\\chromedriver.exe')
driver = webdriver.Chrome(service=path)
#打开登录页面
driver.get('file:///D:/selenium/book/login.html')
#在密码框中输入"testdaniu"
driver.find_element(By.CLASS_NAME,"passwd").send_keys("testdaniu")
```

运行代码后,在密码框中输入字符"testdaniu",因为密码是加密形式的,故显示为"●●●●●●●●",如图 4.18 所示。

图 4.18

在上一节的例子中,用户名也有 class 属性"f-text phone-input",因为不能直接使用空格,所以可以取部分 f-text 或 phone-input 进行定位,代码如下:

```
driver.find_element(By.CLASS_NAME,"f-text").send_keys("testdaniu")
driver.find_element(By.CLASS_NAME,"phone-input").send_keys("testdaniu")
```

4.2.4 link_text 定位

link_text 定位是以超链接全部文本作为关键字来定位元素的。下面以 login.html 页面的"上传资料页面"超链接为例进行介绍,如图 4.19 所示。

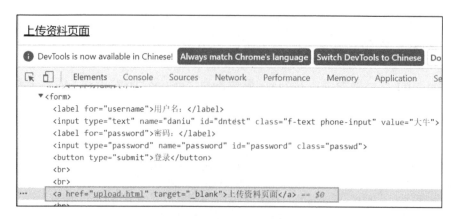

图 4.19

运行以下代码后,浏览器成功地打开了"上传资料页面"超链接,效果如图 4.20 所示。

```
#coding=utf-8
from selenium import webdriver
from selenium.webdriver.chrome.service import Service
from selenium.webdriver.common.by import By
path =Service('D:\\dr\\chromedriver.exe')
driver = webdriver.Chrome(service=path)
#打开登录页面
driver.get('file:///D:/selenium/book/login.html')
#单击"上传资料页面"
driver.find_element(By.LINK_TEXT,"上传资料页面").click()
```

图 4.20

注意:使用 link_text 定位元素超链接,需要写全中文文字。

4.2.5 partial_link_text 定位

partial_link_text 定位使用超链接部分文本来定位元素,类似数据库中的模糊查询。以"上传资料页面"超链接为例,取文本"上传资料"便可定位,即取超链接全部文本的一个子集。示例代码如下:

```
#coding=utf-8
from selenium import webdriver
from selenium.webdriver.chrome.service import Service
from selenium.webdriver.common.by import By
path =Service('D:\\dr\\chromedriver.exe')
driver = webdriver.Chrome(service=path)
#打开登录页面
driver.get('file:///D:/selenium/book/login.html')
#单击"上传资料页面"
driver.find_element(By.PARTIAL_LINK_TEXT,"上传资料").click()
```

4.2.6 CSS 定位

CSS 定位的优点是速度快、语法简单。表 4.1 中的内容出自 W3School 的 CSS 参考手册。CSS 定位的选择器有十几种,在本节中会介绍几种比较常用的选择器。

表 4.1

选 择 器	例 子	例 子 描 述
.class	.intro	选择 class="intro"的所有元素
#id	#firstname	选择 id="firstname"的所有元素
*	*	选择所有元素
element	p	选择所有<p>元素
element,element	div,p	选择所有<div>元素和所有<p>元素
element element	div p	选择<div>元素内部的所有<p>元素
element>element	div>p	选择父元素为<div>元素的所有<p>元素
element+element	div+p	选择紧接在<div>元素之后的所有<p>元素
[attribute]	[target]	选择带有 target 属性的所有元素
[attribute=value]	[target=_blank]	选择 target="_blank"的所有元素
[attribute~=value]	[title~=flower]	选择 title 属性中包含单词"flower"的所有元素

以 class 选择器为例,实现在用户名文本框中输入"大牛测试",代码如下:

```
#coding=utf-8
from selenium import webdriver
from selenium.webdriver.chrome.service import Service
from selenium.webdriver.common.by import By
path =Service('D:\\dr\\chromedriver.exe')
driver = webdriver.Chrome(service=path)
#打开登录页面
driver.get('file:///D:/selenium/book/login.html')
#把元素属性 text 与 phone 之间的空格改成"."
driver.find_element(By.CSS_SELECTOR,".f-text.phone-input").send_keys("大牛测试")
#用.f-text 也可以
driver.find_element(By.CSS_SELECTOR,".f-text").send_keys("大牛测试")
#使用 class='f-text phone-input',不需要处理 text 与 phone 之间的空格
driver.find_element(By.CSS_SELECTOR,"[class='f-text  phone-input']").send_keys("大牛测试")
```

由上可知,id 定位的语法为#加 id 名。实现在用户名文本框中输入"大牛测试",代码如下:

```
#coding=utf-8
from selenium import webdriver
from selenium.webdriver.chrome.service import Service
from selenium.webdriver.common.by import By
path =Service('D:\\dr\\chromedriver.exe')
driver = webdriver.Chrome(service=path)
#打开登录页面
driver.get('file:///D:/selenium/book/login.html')
#在用户名文本框中输入"大牛测试"
driver.find_element(By.CSS_SELECTOR,"#dntest").send_keys("大牛测试")
```

通过常规的标签名称来定位,如用户名文本框的标签为"input",在标签内部又设置了属性值为"name='wd'",测试代码如下:

```
#coding=utf-8
from selenium import webdriver
from selenium.webdriver.chrome.service import Service
from selenium.webdriver.common.by import By
path =Service('D:\\dr\\chromedriver.exe')
```

```
driver = webdriver.Chrome(service=path)
#打开登录页面
driver.get('file:///D:/selenium/book/login.html')
#在用户名文本框中输入"大牛测试"
driver.find_element(By.CSS_SELECTOR,"input[name='daniu']").send_keys("大牛测试")
```

4.2.7 XPath 定位

通过 XPath 定位元素的方式，对比较难以定位的元素来说很有效，几乎都可以实现定位，特别是对于有些元素没有 id、name 等属性的情况。

1. XPath 简介

XPath 是 XML Path 的缩写，是一种用来确定 XML 文件中某部分位置的语言。XPath 在 XML 文件中通过元素名和属性进行搜索，主要用途是在 XML 文件中寻找节点。XPath 定位比 CSS 定位有更大的灵活性。XPath 可以向前搜索也可以向后搜索，而 CSS 定位只能向前搜索，但是 XPath 定位的速度比 CSS 定位的慢一些。

XPath 语言包含根节点、元素、属性、文本、处理指令、命名空间等。以下文本为 XML 实例文件，用于演示 XML 的各种节点类型，便于理解 XPath：

```
<?xml version = "1.0" encoding = "utf-8" ?>
<!-- 这是一个注释节点 -->
<animalList type="mammal">
    <animal categoruy = "forest">
        <name>Tiger</name>
        <size>big</size>
        <action>run</action>
    </animal>
</animalList>
```

其中<animalList>为文件节点，也是根节点；<name>为元素节点；type="mammal"为属性节点。

节点之间的关系介绍如下。

- 父节点。每个元素都有一个父节点，如在上面的 XML 示例中，animal 元素是 name、size 和 action 元素的父节点。

- 子节点。与父节点相反，这里不再赘述。
- 兄弟节点，有时也叫同胞节点。它表示拥有相同父节点的节点。如上面的代码所示，name、size 和 action 元素都是兄弟节点。
- 先辈节点。它是指某节点的父节点或者父节点的父节点，以此类推。如上面的代码所示，name 元素节点的先辈节点有 animal 和 animalList 元素。
- 后代节点。它表示某节点的子节点、子节点的子节点，以此类推。如上面的代码所示，animalList 元素节点的后代节点有 animal、name 等元素。

2. XPath 语法

XPath 来自 XML，又由于 HTML 语言的语法和 XML 的比较接近，故 XPath 也支持定位 HTML 页面元素。下面以定位 login.html 页面登录框为例，采用绝对路径与相对路径进行演示。网页元素如图 4.21 所示。

图 4.21

（1）绝对路径即完整的路径，如 login.html 页面登录框的绝对路径为"html/body/form/input[1]"，完整代码如下：

```
#coding=utf-8
from selenium import webdriver
from selenium.webdriver.chrome.service import Service
from selenium.webdriver.common.by import By
path =Service('D:\\dr\\chromedriver.exe')
driver = webdriver.Chrome(service=path)
driver.get('file:///Users/tim/Desktop/selenium/book/login.html')
```

```
driver.find_element(By.XPATH,"html/body/form/input[1]").send_keys("大牛测试")
```

（2）相对路径方式，以//开头，从当前节点开始解析，"＊"表示通配符，可匹配任何元素节点，"."为选取当前节点，".."为选取当前节点的父节点。实现在 login.html 页面登录框中输入"测试"的代码如下：

```
#coding=utf-8
from selenium import webdriver
from selenium.webdriver.chrome.service import Service
from selenium.webdriver.common.by import By
path =Service('D:\\dr\\chromedriver.exe')
driver = webdriver.Chrome(service=path)
driver.get('file:///D:/selenium/book/login.html')
driver.find_element(By.XPATH,"//*[@id='dntest']").send_keys("测试")
#登录用户名标签是"input"
driver.find_element(By.XPATH,"//input[@id='dntest']").send_keys("测试")
#多个属性条件之间可用 and
driver.find_element(By.XPATH,"//*[@id='dntest'and @name='daniu']").send_keys("测试")
#or 表示只需一个属性条件满足即可
driver.find_element(By.XPATH,"//*[@id='dntest' or @name='daniu']").send_keys("测试")
```

除了以上常规方法定位，部分复杂控件需要采用模糊定位，即使用 starts-with 与 contains。starts-with 表示从某个属性值开始，以 login.html 页面的用户名为例，id 为 dntest，可取部分 id（如 dnt），代码如下：

```
driver.find_element(By.XPATH,"//input[starts-with(@id,'dnt')]").send_keys("大牛测试")
```

contains 表示包含某个属性值，以 login.html 页面的用户名为例，可取部分 id，如 nte，代码如下：

```
driver.find_element(By.XPATH,"//input[contains(@id,'nte')]").send_keys("大牛测试")
```

4.2.8 tag_name 定位

tag_name 定位即通过标签名称定位，常用于复合定位，如图 4.22 所示，定位标签"form"并打印标签属性值"name"。

图 4.22

代码如下：

```
#coding=utf-8
from selenium import webdriver
from selenium.webdriver.chrome.service import Service
from selenium.webdriver.common.by import By
path =Service('D:\\dr\\chromedriver.exe')
driver = webdriver.Chrome(service=path)
driver.get('file:///D:/selenium/book/login.html')
print(driver.find_element(By.TAG_NAME,"form").get_attribute("name"))
```

代码运行成功后，在控制台上打印"daniu"，如图 4.23 所示。

```
/usr/local/bin/python3 /Users/tim/Documents/pyse2023/chapter04/dntest_4.2.8.py
daniu

Process finished with exit code 0
```

图 4.23

4.3　表格定位

在自动化测试过程中，我们可能会碰到要处理各种页面表格的情况，如操作表格中的行、列，以及某些特定的单元格等，所以掌握表格定位的方法也是非常重要的。

4.3.1 遍历表格单元格

以 table1.html 为例，页面内容显示如图 4.24 所示。

职员工资单	201811	201812
张三	10000	12000
王二	5000	5500
麻子	20000	20000
佩奇	1000	2000
小丸子	30000	38000

图 4.24

用循环方式遍历表格中的所有单元格，其代码如下：

```python
from selenium import webdriver
from selenium.webdriver.chrome.service import Service
from selenium.webdriver.common.by import By
#这是在macOS系统上执行的基于Chrome浏览器的测试，如下可根据实际情况修改driver地址
chrome_driver_server = Service("/Users/jason118/Downloads/chromedriver")
driver = webdriver.Chrome(service=chrome_driver_server)
#如下是打开本地文件的代码，可根据实际地址进行改写
driver.get("file:///Users/xxx/PycharmProjects/selenium4.0-automation/Chapter4/table1.html")
#通过id定位获取整个表格对象
html_table = driver.find_element(By.ID,"table1")
#通过元素名（标签）获取表格中的所有行对象
tr_list = html_table.find_elements(By.TAG_NAME,"tr")
th_cols = tr_list[0].find_elements(By.TAG_NAME,"th")
for col in th_cols:
    print(col.text+"\t",end='')
print()
for i in range(1,len(tr_list)):
    td_cols = tr_list[i].find_elements(By.TAG_NAME,"td")
    for c in td_cols:
        print(c.text+"\t",end='')
    print()
```

代码的执行结果如图 4.25 所示，已经正确定位并显示了表格单元格内容。

```
(selenium4.0-automation) (base) 192:Chapter4    $ python table1.py
职员工资单      201811  201812
张三    10000   12000
王二    5000    5500
麻子    20000   20000
佩奇    1000    2000
小丸子   30000   38000
```

图 4.25

代码解释如下：首先通过 id 属性值获取整个表格的页面元素对象。通过元素名获取所有的 tr 对象，最后返回 tr 对象列表。然后，处理 tr 对象列表中的第一个对象，即"tr_list[0]"，因为其是表格的标题行，所以单独处理。

接着，循环遍历第 1 行的每一列，最后返回单元格的文本值，代码如下：

```
th_cols = tr_list[0].find_elements(By.TAG_NAME,"th")
for col in th_cols:
    print(col.text+"\t",end='')
print()
```

循环遍历剩余的行对象，并嵌套循环每一列，最后返回单元格的文本值，代码如下：

```
for i in range(1,len(tr_list)):
    td_cols = tr_list[i].find_elements(By.TAG_NAME,"td")
    for c in td_cols:
        print(c.text+"\t",end='')
    print()
```

4.3.2 定位表格中的特定元素

如果要在被测试网页的表格中定位第 3 行第 3 列的单元格，XPath 定位表达式为：

```
//*[@id="table1"]/tbody/tr[3]/td[3]
```

对应的 Python 定位语句是：

```
driver.find_element(By.XPATH, '//*[@id="table1"]/tbody/tr[3]/td[3]')
```

在表达式中，tr[3]表示第 3 行，td[3]表示第 3 列。从这个层面来说，已经实现了我们的需求。

如果使用 CSS 定位，则表达式为：

```
#table1 > tbody > tr:nth-child(3) > td:nth-child(3)
```

对应的 Python 定位语句是：

```
driver.find_element("By.CSS_SELECTOR","#table1 > tbody > tr:nth-child(3) > td:nth-child(3)")
```

其中 tr:nth-child(3)表示表格的第 3 行，td:nth-child(3)表示表格的第 3 列，这种定位方法也实现了我们的需求。

4.3.3 定位表格中的子元素

以 table2.html 为例，页面内容渲染如图 4.26 所示。

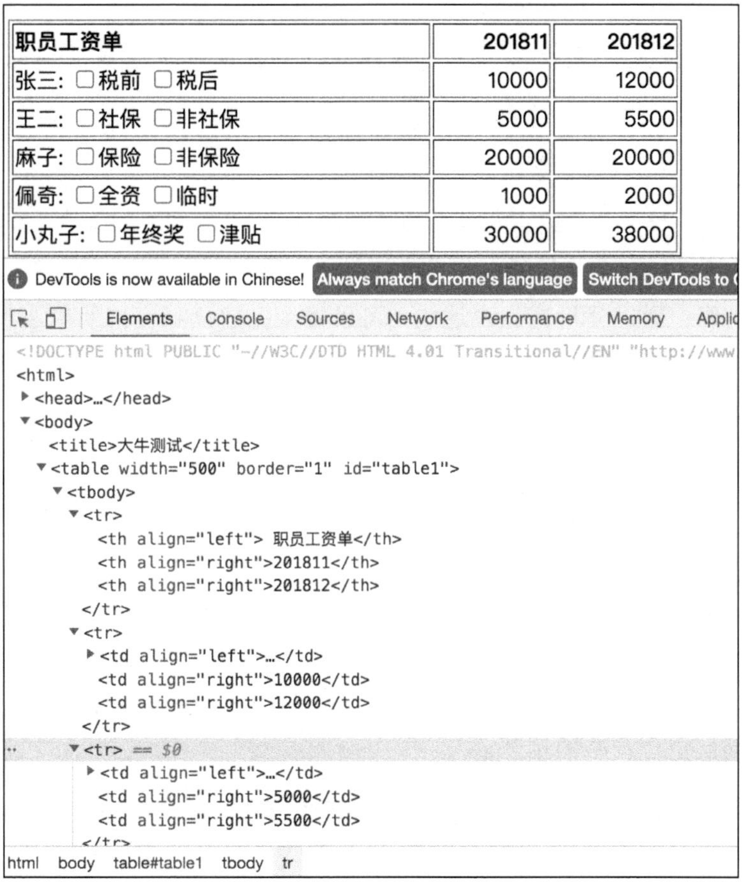

图 4.26

如果我们的需求是在被测试网页中定位表格第 3 行中的"社保"文字前的复选框,则 XPath 定位表达式为:

```
//*[@id="table1"]/tbody/tr[3]/td[1]/input[1]
```

对应的 Python 定位语句是:

```
driver.find_element(By.XPATH, '//*[@id="table1"]/tbody/tr[3]/td[1]/input[1]')
```

定位语句解释:先查找要定位元素所在的单元格,其中"tr[3]/td[1]"表明第 3 行第 1 列,而"input[1]"表明在该单元格内定位第一个 input 元素。

如果使用 CSS 定位,则表达式为:

```
#table1 > tbody > tr:nth-child(3) > td:nth-child(1) > input[type="checkbox"]:nth-child(1)
```

对应的 Python 定位语句是:

```
driver.find_element_by_css_selector('#table1 > tbody > tr:nth-child(3) > td:nth-child(1) > input[type="checkbox"]:nth-child(1)')
```

定位语句解释:通过分析定位表达式,可以看到先定位到表格的第 3 行第 1 列,然后定位单元格内第一个 checkbox 类型的子元素。

4.4 关联元素定位策略

Selenium 4 引入了关联元素定位(Relative Locator)策略。这一策略主要用于应对一些不好定位的元素,但是其周边相关联的元素比较好定位。实现步骤是先定位想要定位的元素周边较容易定位的元素,再根据关联元素定位策略定位想要定位的元素。如下以具体案例讲解关联元素定位策略的用法。

下面以 relative_locator1.html 页面为例,用于后续测试。渲染页面,如图 4.27 所示。

图 4.27

4.4.1 Above 模式

假定"输入用户名"input 元素不好定位,而"输入密码"input 元素容易定位,此时就可以利用关联元素定位策略的 Above 模式来定位"输入用户名"input 元素。

下面实现通过 password input 元素获取 username input 元素,并且在 username 输入框中输入字符"name1",代码如下:

```
from selenium import webdriver
from selenium.webdriver.chrome.service import Service
from selenium.webdriver.common.by import By
from selenium.webdriver.support.relative_locator import locate_with
#这是在 macOS 系统上执行的基于 Chrome 浏览器的测试,如下可根据实际情况修改 driver 地址
```

```
chrome_driver_server = Service("/Users/xxx/Downloads/chromedriver")
driver  = webdriver.Chrome(service=chrome_driver_server)
#如下是打开本地文件的代码，可根据实际地址进行改写
driver.get("file:///Users/jason118/PycharmProjects/selenium4.0-automation/Chapter4/relative_locator1.html")
#通过关联元素定位策略的Above模式先获取password input，然后获取username input
username_locator = locate_with(By.TAG_NAME,"input").above({By.ID: "id_pass"})
username = driver.find_element(username_locator)
username.send_keys('name1')
```

代码解释如下：

- 使用关联元素定位策略时，需要引入 locate_with，引入语句为"from selenium.webdriver.support.relative_locator import locate_with"。

- 通过 Above 模式获取 relative locator 对象。

- 将 relative locator 对象以参数的形式传入方法 find_element。

代码执行结果如图 4.28 所示。

图 4.28

4.4.2 Below 模式

以上面的 HTML 页面为例，假设"输入密码"input 元素不好定位，而"输入用户名"

input 元素容易定位，则可利用关联元素定位策略的 Below 模式来定位"输入密码"input 元素。

下面实现通过 username input 元素获取 password input 元素，并且在 password 输入框中输入字符"password1"，代码如下：

```python
from selenium import webdriver
from selenium.webdriver.chrome.service import Service
from selenium.webdriver.common.by import By
from selenium.webdriver.support.relative_locator import locate_with
#这是在macOS 系统上执行的基于Chrome 浏览器的测试，如下可根据实际情况修改driver 地址
chrome_driver_server = Service("/Users/xxx/Downloads/chromedriver")
driver = webdriver.Chrome(service=chrome_driver_server)
#如下是打开本地文件的代码，可根据实际地址进行改写
driver.get("file:///Users/jason118/PycharmProjects/selenium4.0-automation/Chapter4/relative_locator1.html")
#通过关联元素定位策略的Below 模式先获取username input，然后获取password input
password_locator = locate_with(By.TAG_NAME,"input").below({By.ID: "id_username"})
password = driver.find_element(password_locator)
password.send_keys('daniu')
```

以上代码与 Above 模式代码的不同点是，将代码中的 Above 模式改为了 Below 模式。

- Above 模式使用方式为"locate_with(By.TAG_NAME,"input").above"。
- Below 模式使用方式为"locate_with(By.TAG_NAME,"input").below"。

代码执行结果如图 4.29 所示。

图 4.29

4.4.3　Left of 模式

以上面的 HTML 页面为例，假设"取消"按钮不好定位，而右边的"登录"按钮容易定位，则可利用关联元素定位策略的 Left of 模式来定位"取消"按钮元素，代码如下：

```
from selenium import webdriver
from selenium.webdriver.chrome.service import Service
from selenium.webdriver.common.by import By
from selenium.webdriver.support.relative_locator import locate_with
#这是在macOS系统上执行的基于Chrome浏览器的测试，如下可根据实际情况修改driver地址
chrome_driver_server = Service("/Users/xxx/Downloads/chromedriver")
driver  = webdriver.Chrome(service=chrome_driver_server)
#如下是打开本地文件的代码，可根据实际地址进行改写
driver.get("file:///Users/jason118/PycharmProjects/pyse2023 /Chapter4/relative_locator1.html")
#通过关联元素定位策略的Left of模式先获取"登录"按钮，然后获取"取消"按钮
cancel_locator = locate_with(By.TAG_NAME,"button").to_left_of({By.ID: "id_login"})
cancel_element = driver.find_element(cancel_locator)
#输出"取消"按钮元素
print(cancel_element)
```

4.4.4　Right of 模式

以上面的 HTML 页面为例，假设"登录"按钮不好定位，而左边的"取消"按钮容易定位，则可利用关联元素定位策略的 Right of 模式来定位"登录"按钮元素，代码如下：

```
from selenium import webdriver
from selenium.webdriver.chrome.service import Service
from selenium.webdriver.common.by import By
from selenium.webdriver.support.relative_locator import locate_with
#这是在macOS系统上执行的基于Chrome浏览器的测试，如下可根据实际情况修改driver地址
chrome_driver_server = Service("/Users/xxx/Downloads/chromedriver")
driver  = webdriver.Chrome(service=chrome_driver_server)
#如下是打开本地文件的代码，可根据实际地址进行改写
driver.get("file:///Users/jason118/PycharmProjects/pyse2023 /Chapter4/relative_locator1.html")
#通过关联元素定位策略的Right of模式先获取"取消"按钮，然后获取"登录"按钮
```

```
login_locator = locate_with(By.TAG_NAME,"button").to_right_of({By.ID: "id_cancel"})
login_element = driver.find_element(login_locator)
#输出"登录"按钮元素
print(login_element)
```

4.4.5　Near 模式

对于某些元素与元素之间的相对位置关系不是很明确的情况，如元素 A 不在元素 B 的正上方、正下方、正右边、正左边等时，可采用 Near 模式，即目标元素在某元素的附近（方圆 50px 之内）也可被定位到。

以上面的 HTML 页面为例，如果要定位"输入用户名："label 元素，可以先定位"输入用户名"input 元素，再利用关联元素定位策略的 Near 模式来定位"输入用户名："label 元素，代码如下：

```
from selenium import webdriver
from selenium.webdriver.chrome.service import Service
from selenium.webdriver.common.by import By
from selenium.webdriver.support.relative_locator import locate_with
#这是在 macOS 系统上执行的基于 Chrome 浏览器的测试，如下可根据实际情况修改 driver 地址
chrome_driver_server = Service("/Users/jason118/Downloads/chromedriver")
driver = webdriver.Chrome(service=chrome_driver_server)
#如下是打开本地文件的代码，可根据实际地址进行改写
driver.get("file:///Users/jason118/PycharmProjects/selenium4.0-automation/Chapter4/relative_locator1.html")
label_username_locator = locate_with(By.TAG_NAME,"label").near({By.ID:"id_username"})
label_username_element = driver.find_element(label_username_locator)
print(label_username_element)
```

4.4.6　Chaining relative locators 模式

Chaining relative locators 模式的意思举例来说就是，目标元素的位置既满足在元素 A 的 Above 位置，又满足在元素 B 的 Right of 位置。

以上面的 HTML 页面为例，假设"取消"按钮元素不好定位，则可利用关联元素定位策略的 Chaining relative locators 模式进行定位。"取消"按钮元素需要既满足"输入密

码"label 元素的 Below 位置，又满足"登录"按钮元素的 Left of 位置，代码如下：

```
from selenium import webdriver
from selenium.webdriver.chrome.service import Service
from selenium.webdriver.common.by import By
from selenium.webdriver.support.relative_locator import locate_with
#这是在macOS系统上执行的基于Chrome浏览器的测试，如下可根据实际情况修改driver地址
chrome_driver_server = Service("/Users/jason118/Downloads/chromedriver")
driver  = webdriver.Chrome(service=chrome_driver_server)
#如下是打开本地文件的代码，可根据实际地址进行改写
driver.get("file:///Users/jason118/PycharmProjects/pyse2023 /Chapter4/relative_locator1.html")
cancel_button_locator = locate_with(By.TAG_NAME,"button").below({By.ID:"id_label2"}).to_left_of({By.ID: "id_login"})
cancel_button_element = driver.find_element(cancel_button_locator)
#输出元素对象
print(cancel_button_element)
```

本章主要介绍了 Selenium 元素的八大定位方法，每一种定位方法都有其特殊的用法，读者只要掌握其用法即可。

第 5 章 Selenium 常用方法

Selenium 常用方法与 Selenium 元素定位类似，都是和页面元素打交道。元素定位负责在页面上定位期望元素，常用方法则是对这些元素做出一些期望操作，两者配合才能实现 UI 自动化测试。

5.1 基本方法

1. send_keys 方法

该方法类似于模拟键盘键入。以在 login.html 用户名输入框中输入"测试"为例，代码如下：

```
#coding=utf-8
#导入 WebDriver 模块
from selenium import webdriver
```

```
from selenium.webdriver.chrome.service import Service
from selenium.webdriver.common.by import By
path =Service('D:\\dr\\chromedriver.exe')
driver = webdriver.Chrome(path)
driver.get('file:///D:/selenium/book/login.html')
#执行后，在用户名输入框中输入"测试"
driver.find_element(By.ID,"dntest").send_keys("测试")
```

2. text 方法

Selenium 提供了 text 方法，用于获取文本值，即 HTML 标签<a>和</a>之间的文字，如图 5.1 所示。

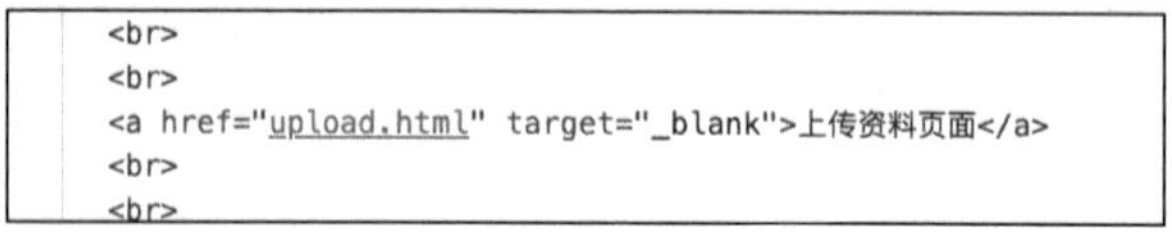

```
<br>
<br>
<a href="upload.html" target="_blank">上传资料页面</a>
<br>
<br>
```

图 5.1

以在 login.html 页面获取超链接"上传资料页面"文字为例，代码如下：

```
#coding=utf-8
from selenium import webdriver
from selenium.webdriver.chrome.service import Service
from selenium.webdriver.common.by import By
path =Service('D:\\dr\\chromedriver')
driver = webdriver.Chrome(path)
driver.get('file:///D:/selenium/book/login.html')
#执行后，在控制台上打印"上传资料页面"
print(driver.find_element(By.LINK_TEXT,"上传资料页面").text)
```

3. get_attribute 方法

以 login.html 用户名输入框为例，若想获取 class 属性值，可以用 get_attribute 方法来实现，页面元素如图 5.2 所示。

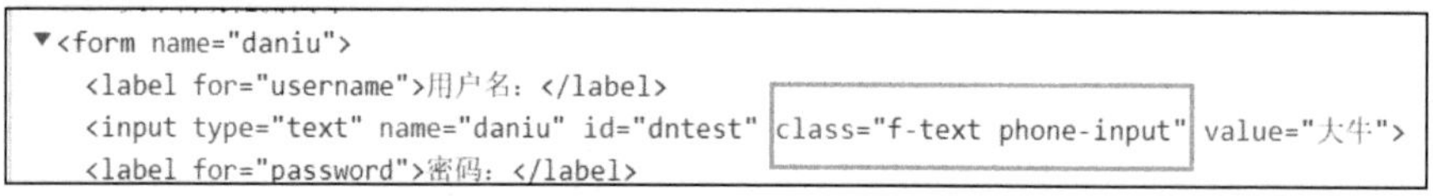

```
▼<form name="daniu">
    <label for="username">用户名：</label>
    <input type="text" name="daniu" id="dntest" class="f-text phone-input" value="大牛">
    <label for="password">密码：</label>
```

图 5.2

示例代码如下：

```
#coding=utf-8
from selenium import webdriver
from selenium.webdriver.chrome.service import Service
from selenium.webdriver.common.by import By
path =Service('D:\\dr\\chromedriver')
driver = webdriver.Chrome(service=path)
driver.get('file:///D:/selenium/book/login.html')
print(driver.find_element(By.LINK_TEXT,"上传资料页面").get_attribute
("class"))
```

注意，有些 Selenium 取值方法效果是一样的，如下两行代码的效果就是一样的，都会输出"上传资料页面"字符串：

```
driver.find_element(By.LINK_TEXT,"上传资料页面").get_attribute
('textContent')
  driver.find_element(By.LINK_TEXT,"上传资料页面").text
```

4. maximize_window 方法

maximize_window 方法用来实现浏览器窗口最大化操作，代码如下：

```
#coding=utf-8
from selenium import webdriver
from selenium.webdriver.chrome.service import Service
path =Service('D:\\dr\\chromedriver.exe')
driver = webdriver.Chrome(path)
driver = webdriver.Chrome()
#浏览器窗口最大化
driver.maximize_window()
```

5. minimize_window 方法

minimize_window 方法用来实现浏览器窗口最小化操作，代码如下：

```
driver.minimize_window()
```

6. fullscreen_window 方法

fullscreen_window 方法用来实现全屏操作,代码如下:

```
driver.fullscreen_window()
```

7. current_window_handle 方法

current_window_handle 方法用于返回窗口句柄,即标识窗口字符串,如图 5.3 所示,当前窗口的句柄字符串是 "CDwindow-179C72265B9C40777D10EAC2238DC3EC"。

```
/usr/local/bin/python3 /Users/tim/Documents/pyse2023/chapter05/dntest_5.1.4.py
CDwindow-179C72265B9C40777D10EAC2238DC3EC

Process finished with exit code 0
```

图 5.3

如下代码输出 login.html 的窗口句柄,需要注意的是,每次运行代码后输出的句柄字符串都不同:

```
from selenium import webdriver
from selenium.webdriver.chrome.service import Service
from selenium.webdriver.common.by import By
path =Service('D:\\dr\\chromedriver.exe')
driver = webdriver.Chrome(service=path)
#打开登录页面
driver.get('file:///D:/selenium/book/login.html')
print(driver.current_window_handle)
```

8. current_url 方法

current_url 方法用来获取当前窗口 URL,比如输出当前窗口的 URL 为"login.html url",代码如下:

```
from selenium import webdriver
from selenium.webdriver.chrome.service import Service
from selenium.webdriver.common.by import By
path =Service('/Users/tim/Downloads/chromedriver ')
driver = webdriver.Chrome(service=path)
#打开登录页面
```

```
driver.get('file:///Users/tim/Desktop/selenium/book/login.html')
print(driver.current_url)
```

运行代码后，结果如图 5.4 所示，当前浏览器窗口的 URL 是 "file:///Users/tim/Desktop/selenium/book/login.html"。

```
/usr/local/bin/python3 /Users/tim/Documents/pyse2023/chapter05/dntest_5.1.4.py
file:///Users/tim/Desktop/selenium/book/login.html

Process finished with exit code 0
```

图 5.4

9. is_selected 方法

is_selected 方法用于判断元素是否被选择，多用于选择框，如果多选框处于被选中的状态，则返回 True，反之则返回 False，以在 login.html 页面上选择男女为例，默认选择女，如图 5.5 所示。

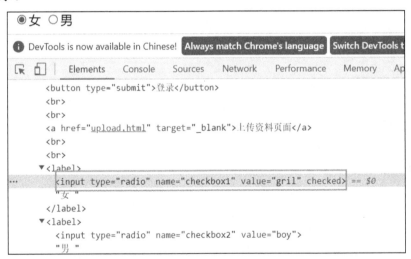

图 5.5

实现男、女选项的默认状态，代码如下：

```
#coding=utf-8
from selenium import webdriver
from selenium.webdriver.chrome.service import Service
path= 'D:/dr/chromedriver.exe'
from selenium.webdriver.common.by import By
```

```
path =Service('D:\\dr\\chromedriver')
driver = webdriver.Chrome(service=path)
driver.get('file:///D:/selenium/book/login.html')
print(driver.find_element(By.NAME,"checkbox1").is_selected())
print(driver.find_element(By.NAME,"checkbox2").is_selected())
```

10. is_enabled 方法

is_enabled 方法用于判断页面元素是否可用,可用则返回 True,不可用则返回 False。如 login.html 用户名为可用状态,则返回 True,示例代码如下:

```
print(driver.find_element(By.ID,"dntest").is_enabled)
```

11. is_displayed 方法

is_displayed 方法用于判断页面中是否显示元素,显示则返回 True,不显示则返回 False。如 login.html 用户名为显示状态,则返回 True,示例代码如下:

```
print(driver.find_element(By.ID,"dntest").is_displayed)
```

12. clear 方法

clear 方法用于清除输入框内的值。如 login.html 用户名输入框中有默认值"大牛",实际项目中需要先清除默认值再输入新值,示例代码如下:

```
#coding=utf-8
from selenium import webdriver
from selenium.webdriver.chrome.service import Service
from selenium.webdriver.common.by import By
path =Service('D:\\dr\\chromedriver.exe')
driver = webdriver.Chrome(service=path)
driver.get('file:///D:/selenium/book/login.html')
driver.find_element(By.ID,"dntest").clear()
```

13. quit 方法

quit 方法用于关闭浏览器并"杀掉"chromedriver.exe 进程。以 Windows 为例,运行该方法后任务管理器中的驱动进程将被"杀掉",如图 5.6 所示。

第 5 章 Selenium 常用方法

图 5.6

14. title 方法

title 方法用于获取页面 title。以 login.html 页面为例，对应的 title 为 "大牛测试"，示例代码如下：

```
#coding=utf-8
from selenium import webdriver
from selenium.webdriver.chrome.service import Service
path =Service('D:\\dr\\chromedriver.exe')
driver = webdriver.Chrome(service=path)
driver.get('file:///D:/selenium/book/login.html')
#在控制台上打印title "大牛测试"
print(driver.title)
```

15. refresh 方法

refresh 方法用于刷新页面，类似键盘上的 F5 键或 Ctrl+F5 快捷键，示例代码如下：

```
#coding=utf-8
from selenium import webdriver
from selenium.webdriver.chrome.service import Service
path =Service('D:\\dr\\chromedriver')
```

```
driver = webdriver.Chrome(service=path)
driver.get('file:///D:/selenium/book/login.html')
#刷新当前页面
driver.refresh()
```

16. back 方法

back 方法用于浏览器工具栏向后操作，以访问 login.html 页面后退至空页面为例，示例代码如下：

```
#coding=utf-8
from selenium import webdriver
from selenium.webdriver.chrome.service import Service
path =Service('D:\\dr\\chromedriver')
driver = webdriver.Chrome(service=path)
driver.get('file:///D:/selenium/book/login.html')
#浏览器工具栏向后操作
driver.back();
```

17. forward 方法

forward 方法用于浏览器工具栏向前操作，示例代码如下：

```
#coding=utf-8
from selenium import webdriver
from selenium.webdriver.chrome.service import Service
path =Service('D:\\dr\\chromedriver')
driver = webdriver.Chrome(service=path)
driver.get('file:///D:/selenium/book/login.html')
#浏览器工具栏向后操作
driver.back();
#浏览器工具栏向前操作
driver.forward();
```

18. name 方法

name 方法用于返回当前运行的浏览器名称，示例代码如下：

```
#coding=utf-8
```

```
from selenium import webdriver
from selenium.webdriver.chrome.service import Service
path =Service('D:\\dr\\chromedriver')
driver = webdriver.Chrome(service=path)
#打开登录页面
driver.get('file:///D:/selenium/book/login.html')
print(driver.name)
```

19. save_screenshot 方法

save_screenshot 方法用于对页面截图，常用于执行失败时对页面截图，便于定位错误，以 login.html 页面截图为例，代码运行后在脚本的当前目录下生成文件名为"picture.png"的图片，示例代码如下：

```
from selenium import webdriver
from selenium.webdriver.chrome.service import Service
#driver 地址
path =Service('D:\\dr\\chromedriver.exe')
driver = webdriver.Chrome(service=path)
driver.get("file:///D:/selenium/book/login.html ")
driver.maximize_window()
#截图操作，图片文件名为"picture.png"
driver.save_screenshot("picture.png")
```

20. get_screenshot_as_base64 方法

get_screenshot_as_base64 方法用于对窗口截图并以 BASE64 方式返回截图，如以对 login.html 页面截图打印编码为例，示例代码如下：

```
#coding=utf-8
from selenium import webdriver
from selenium.webdriver.chrome.service import Service
from selenium.webdriver.common.by import By
path =Service('D:\\dr\\chromedriver')
driver = webdriver.Chrome(service=path)
#打开登录页面
driver.get('file:///D:/selenium/book/login.html')
print(driver.get_screenshot_as_base64())
```

把编码复制到编码转换工具中,将编码转为图片,如图 5.7 所示。

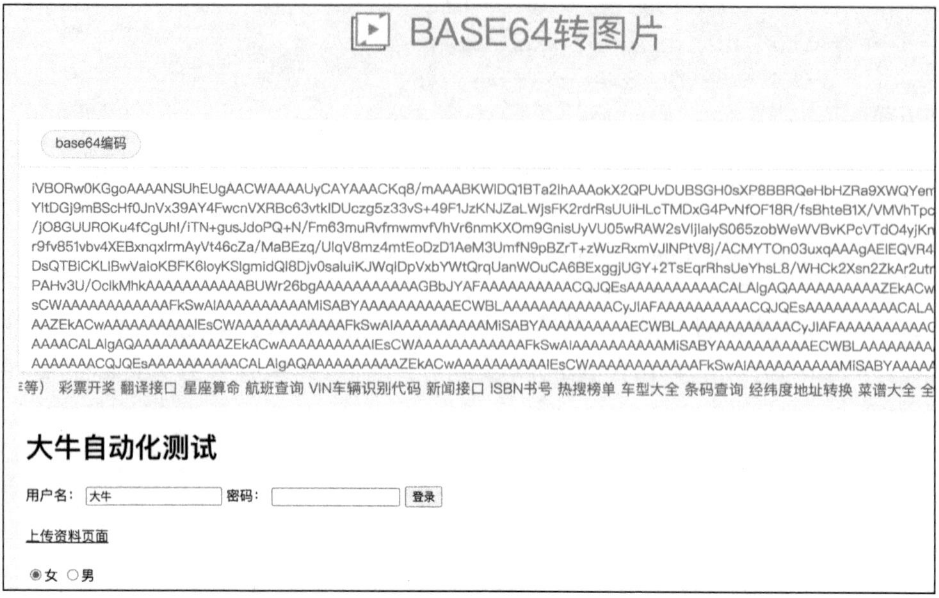

图 5.7

21. page_source 方法

page_source 方法用于输出网页源码,如 login.html 页面源码,示例代码如下:

```
#coding=utf-8
from selenium import webdriver
from selenium.webdriver.chrome.service import Service
from selenium.webdriver.common.by import By
path =Service('D:\\dr\\chromedriver')
driver = webdriver.Chrome(service=path)
driver.get('file:///D:/selenium/book/login.html')
print(driver.page_source)
```

22. set_window_size 方法

set_window_size 方法用于设置浏览器当前窗口大小,设置窗口宽度为 800、高度为 600,示例代码如下:

```
driver.set_window_size(800,600)
```

23. get_window_size 方法

get_window_size 方法用于获取浏览器当前窗口的高度与宽度，示例代码如下：

```
driver.get_window_size()
```

24. set_window_position 方法

set_window_position 方法用于设置浏览器当前窗口的位置，设置 x,y 为 0,0，示例代码如下：

```
driver.set_window_position(0,0)
```

25. click 方法

click 方法用于实现单击操作，如 button 登录操作，以 login.html 页面登录为例，示例代码如下：

```
driver.find_element(By.ID,"loginbtn").click()
```

26. window_handles 方法

window_handles 方法用于获取多窗口句柄，返回列表类型，如打开 login.html 页面，单击"上传资料页面"后，输出所有窗口句柄，如图 5.8 所示，代码如下：

```
#coding=utf-8
from selenium import webdriver
from selenium.webdriver.common.by import By
path = "/Users/tim/Downloads/chromedriver"
driver = webdriver.Chrome(path)
driver.get('file:///Users/tim/Desktop/selenium/book/login.html')
#打印当前窗口句柄
driver.find_element(By.PARTIAL_LINK_TEXT,"上传资料").click()
print(driver.window_handles)
```

```
/usr/local/bin/python3.11 /Users/tim/Documents/pyse2023/chapter05/dntest_5.1.27.py
['CDwindow-6013C69F2EF31A5E50EB66F4413E55DA', 'CDwindow-2786CACE1ACCCDF681AAD94F0CCF1BC3']

Process finished with exit code 0
```

图 5.8

5.2 特殊元素定位

5.2.1 鼠标事件操作

在 Selenium 中，将键盘鼠标操作封装在了 Action Chains 类中。Action Chains 类的主要应用场景为单击鼠标、双击鼠标、鼠标拖曳等。部分常用方法介绍如下。

- click(on_element=None)，模拟鼠标单击操作。
- click_and_hold(on_element=None)，模拟鼠标单击操作并且按住不放。
- double_click(on_element=None)，模拟鼠标双击操作。
- context_click(on_element=None)，模拟鼠标右击操作。
- drag_and_drop(source,target)，模拟鼠标拖曳。
- drag_and_drop(source,xoffset,yoffset)，模拟将目标拖曳到目标位置。
- key_down(value,element=None)，模拟按住某个键，实现快捷键操作。
- key_up(value,element=None)，模拟松开某个键，一般和 key_down 方法一起使用。
- move_to_element(to_element)，模拟将鼠标移到指定的某个页面元素上。
- move_to_element_with_offset(to_element,xoffset,yoffset)，模拟移动鼠标光标至指定的坐标。
- perform()，执行之前一系列的 Action Chains。
- release(on_element=None)，释放按下的鼠标。

接下来列举鼠标悬停、鼠标右击操作和鼠标双击操作 3 个实例。

鼠标悬停，即当鼠标光标在页面元素上停留时触发的事件，以 hover.html 系统控件为例，如图 5.9 所示。

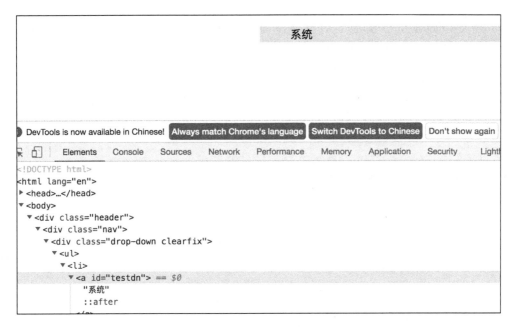

图 5.9

使用 move_to_element 方法，鼠标即可悬停于元素"系统"上，代码如下：

```
from selenium import webdriver
from selenium.webdriver.chrome.service import Service
from selenium.webdriver.common.action_chains import ActionChains
from selenium.webdriver.common.by import By
path= Service('D:/dr/chromedriver.exe')
driver = webdriver.Chrome(service=path)
driver.get('file:///D:/selenium/book/hover.html#')
#鼠标悬停在"系统"元素上
ActionChains(driver).move_to_element(driver.find_element(By.ID,"testdn")).perform()
#单击"登录"按钮
driver.find_element(By.ID,"dntest").click()
driver.quit()
```

使用右键操作方法 context_click，以 login.html "登录"按钮为例，代码如下：

```
#coding=utf-8
from selenium import webdriver
```

```
from selenium.webdriver.chrome.service import Service
from selenium.webdriver.common.by import By
from selenium.webdriver import ActionChains
path =Service('D:\\dr\\chromedriver')
driver = webdriver.Chrome(service=path)
#打开登录页面
driver.get('file:///D:/selenium/book/login.html')
element = driver.find_element(By.ID,"dntest")
ActionChains(driver).context_click(element).perform()
```

执行双击操作，以 login.html 页面的"上传资料页面"为例，代码如下：

```
#coding=utf-8
from selenium import webdriver
from selenium.webdriver import ActionChains
driver = webdriver.Chrome()
driver.maximize_window()
driver.get('file:///D:/selenium/book/login.html')
element = driver.find_element(By.LINK_TEXT,"上传资料页面")
#双击"上传资料页面"
ActionChains(driver).double_click(element).perform()
```

5.2.2 常用的键盘事件

以下为自动化测试中常用的键盘事件。

- Keys.BACK_SPACE：删除键。

- Keys.SPACE：空格键。

- Keys.TAB：Tab 键。

- Keys.ESCAPE：回退键。

- Keys.ENTER：回车键。

- Keys.CONTROL,"a"：快捷键 Ctrl + A。

- Keys.CONTROL,"x"：快捷键 Ctrl + X。

- Keys.CONTROL,"v"：快捷键 Ctrl + V。
- Keys.CONTROL,"c"：快捷键 Ctrl + C。
- Keys.F1：F1 键。
- Keys.F12：F12 键。

用法举例，实现在 login.html 页面登录框中输入文本"大牛测试"并删除输入的最后一个字符，代码如下：

```
#coding=utf-8
from selenium import webdriver
from selenium.webdriver.common.keys import Keys
driver = webdriver.Chrome()
path =Service('D:\\dr\\chromedriver')
driver = webdriver.Chrome(service=path)
#打开登录页面
driver.get('file:///D:/selenium/book/login.html')
#输入"大牛测试"
driver.find_element(By.ID,"dntest").send_keys("大牛测试")
#删除最后一个字"试"
driver.find_element_by_id("kw").send_keys(Keys.BACK_SPACE)
```

注意，最后两行代码也可以进行合并，写法如下：

```
driver.find_element_by_id("kw").send_keys("大牛测试"+Keys.BACK_SPACE)
```

5.2.3 Select 操作

在自动化测试过程中，经常会遇到需要定位处理页面的 Select 元素，而 Selenium 提供了处理 Select 元素的方法。常用方法如下。

- select_by_index
- select_by_value
- select_by_visible_text

接下来具体介绍这 3 种方法，以 select.html 页面为例，如图 5.10 所示。

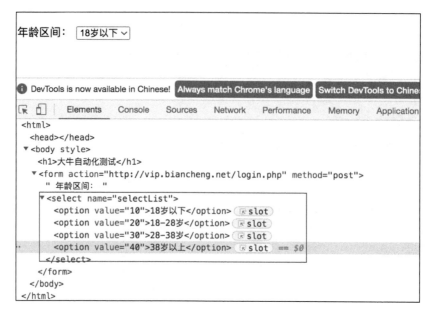

图 5.10

（1）通过 select_by_index 方法获取下拉框选项，代码如下：

```
from selenium.webdriver.support.select import Select
from selenium import webdriver
from selenium.webdriver.chrome.service import Service
from selenium.webdriver.common.by import By
#这是在macOS上执行的基于Chrome浏览器的测试，请根据实际情况修改如下的driver地址
chrome_driver_server = Service("/Users/xxx/Downloads/chromedriver")
driver = webdriver.Chrome(service=chrome_driver_server)
driver.maximize_window()
driver.get("file:///Users/pyse2023/Chapter5/select.html")
select_element = driver.find_element(By.NAME," selectList")
#index索引是从0开始的，如下代码中选择1，表示第2个选项
Select(select_element).select_by_index(1)
```

执行效果如图 5.11 所示，选择了选项"18-28 岁"。

图 5.11

（2）通过 select_by_value 方法获取下拉框选项，代码如下：

```
from selenium.webdriver.support.select import Select
from selenium import webdriver
from selenium.webdriver.chrome.service import Service
from selenium.webdriver.common.by import By
#这是在macOS上执行的基于Chrome浏览器的测试，请根据实际情况修改如下的driver地址
chrome_driver_server = Service("/Users/xxx/Downloads/chromedriver")
driver  = webdriver.Chrome(service=chrome_driver_server)
driver.maximize_window()
driver.get("file:///Users/pyse2023/Chapter5/select.html")
select_element = driver.find_element(By.NAME," selectList")
#value 值为 20 表示第 2 个选项
Select(select_element).select_by_value("20")
#driver.quit()
```

（3）用 select_by_visible_text 方法获取下拉框选项，在本例中 visible text 为 "18 岁以下"、"18-28 岁"、"28-38 岁" 和 "38 岁以上"，如图 5.12 所示。

```
▼<select name="selectList">
    <option value="10">18岁以下</option> slot
    <option value="20">18-28岁</option> slot
    <option value="30">28-38岁</option> slot
    <option value="40">38岁以上</option> slot == $0
  </select>
```

图 5.12

相关代码如下：

```
from selenium.webdriver.support.select import Select
from selenium import webdriver
from selenium.webdriver.chrome.service import Service
from selenium.webdriver.common.by import By
#这是在macOS上执行的基于Chrome浏览器的测试，请根据实际情况修改如下的driver地址
chrome_driver_server = Service("/Users/xxx/Downloads/chromedriver")
driver  = webdriver.Chrome(service=chrome_driver_server)
driver.maximize_window()
driver.get("file:///Users/pyse2023/Chapter5/select.html")
select_element = driver.find_element(By.NAME," selectList")
#visible_text 值为 "18-28 岁"，表示第 2 个选项
Select(select_element).select_by_visible_text("18-28岁")
#driver.quit()
```

以上用封装好的 select 方法对下拉框选项进行操作。对返回选项（options）的信息，也提供了 3 种常用方法，下面分别进行介绍。

（1）options，返回下拉框中的所有选项，代码如下：

```
#coding=utf-8
import time
from selenium import webdriver
from selenium.webdriver.support.select import Select
from selenium.webdriver.common.by import By
from selenium.webdriver.chrome.service import Service
chromedriver = Service("/Users/tim/Downloads/chromedriver")
driver  = webdriver.Chrome(service=chromedriver)
driver.get("file:///Users/pyse2023/Chapter5/select.html")
se = driver.find_element(By.NAME,"selectList")
ops = Select(se).options
for i in ops:
    print(i.text)
driver.quit()
```

运行结果如图 5.13 所示，在控制台上打印了下拉框中的所有选项。

```
/usr/local/bin/python3 /Users/tim/Documents/pyse2023/chapter05/daniu_5.2.3_4.py
18岁以下
18-28岁
28-38岁
38岁以上

Process finished with exit code 0
```

图 5.13

（2）all_selected_options，返回下拉框中已选中的选项，代码如下：

```
#coding=utf-8
import time
from selenium import webdriver
from selenium.webdriver.support.select import Select
from selenium.webdriver.common.by import By
from selenium.webdriver.chrome.service import Service
chromedriver = Service("/Users/tim/Downloads/chromedriver")
driver  = webdriver.Chrome(service=chromedriver)
driver.get("file:///D:/selenium/book/select.html")
```

```
se = driver.find_element(By.NAME,"selectList")
ops = Select(se).all_selected_options
for i in ops:
    print(i.text)
```

运行结果如图 5.14 所示,在控制台上打印了第 1 个选项。

```
/usr/local/bin/python3 /Users/tim/Documents/pyse2023/chapter05/daniu_5.2.3_5.py
18岁以下

Process finished with exit code 0
```

图 5.14

(3) first_selected_option,返回第 1 个被选中的选项,代码如下:

```
#coding=utf-8
import time
from selenium import webdriver
from selenium.webdriver.support.select import Select
from selenium.webdriver.common.by import By
from selenium.webdriver.chrome.service import Service
chromedriver = Service("/Users/tim/Downloads/chromedriver")
driver   = webdriver.Chrome(service=chromedriver)
driver.get("file:///D:/selenium/book/select.html")
se = driver.find_element(By.NAME,"selectList")
ops = Select(se).first_selected_option
print(ops.text)
```

运行结果如图 5.15 所示,在控制台上打印了第 1 个被选中的选项。

```
/usr/local/bin/python3 /Users/tim/Documents/pyse2023/chapter05/daniu_5.2.3_6.py
18岁以下

Process finished with exit code 0
```

图 5.15

5.2.4 定位一组元素

八大定位中都是定位单个元素,八大定位也包含 8 种复合定位,定位单个元素的方法为 find_element,定位一组元素的方法为 find_elements,以 checkbox.html 页面为例,如图 5.16 所示。

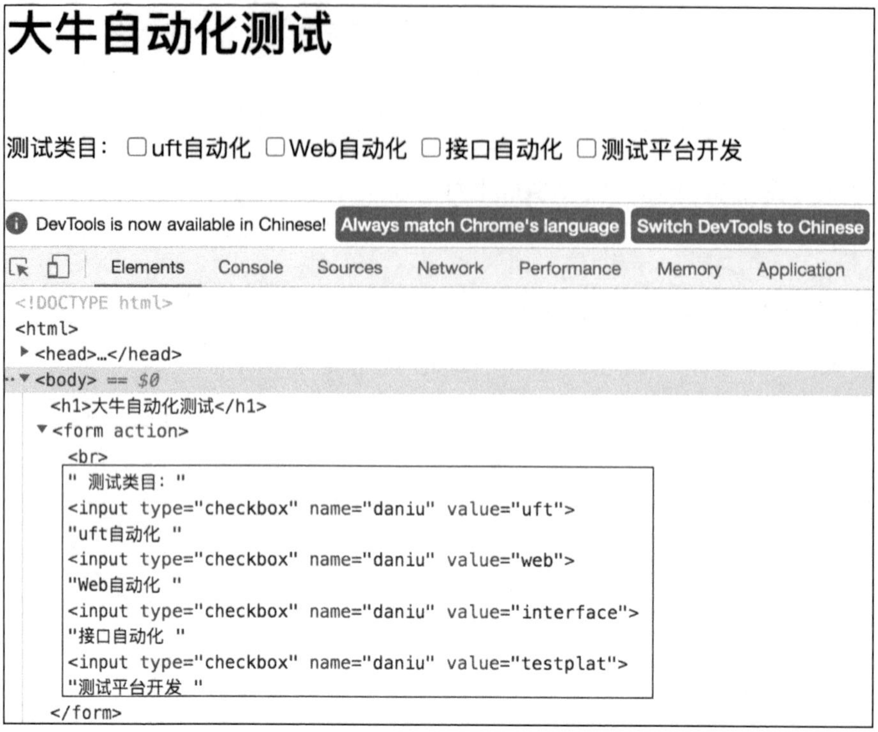

图 5.16

实现选中"Web 自动化"框,用复合定位方法实现,代码如下:

```
#coding=utf-8
from selenium import webdriver
from selenium.webdriver.chrome.service import Service
#复合定位
path= 'D:/dr/chromedriver.exe'
from selenium.webdriver.common.by import By
path =Service('D:\\dr\\chromedriver.exe')
driver = webdriver.Chrome(service=path)
driver.get('file:///D:/selenium/book/checkbox.html')
elements = driver.find_elements(By.NAME,"daniu")
#打印所有NAME为"daniu"的元素,存放到列表中
print(elements)
#单击第2个元素,即"Web自动化"框
elements[1].click()
```

5.3 Frame 操作

Frame 标签有 Frameset、Frame 和 iFrame 这 3 种。Frameset 可以直接按照正常元素定位。Frame 和 iFrame 的定位方法相同，需要把驱动切换到 Frame 内进行操作。下面以登录 frame.html 页面为例来进行讲解，如图 5.17 所示。

图 5.17

第一步，我们尝试用常规方式对页面进行操作，代码如下：

```
from selenium import webdriver
from selenium.webdriver.chrome.service import Service
from selenium.webdriver.common.by import By
path =Service('D:\\dr\\chromedriver')
driver = webdriver.Chrome(service=path)
driver.get('file:///D:/selenium/book/frame.html')
#driver.switch_to.frame("frame")
driver.find_element(By.ID,"dntest").send_keys("test")
```

代码执行后操作失败,并提示找不到元素,错误信息如下:

```
selenium.common.exceptions.NoSuchElementException: Message: no such element: Unable to locate element: {"method":"css selector","selector":"[id="dntest"]"}
```

第二步,按照 Frame 元素定位模式来定位,iFrame 的 id 属性值为"frame",先进行驱动切换操作,代码如下:

```
#coding=utf-8
from selenium import webdriver
from selenium.webdriver.chrome.service import Service
from selenium.webdriver.common.by import By
path =Service('D:\\dr\\chromedriver')
driver = webdriver.Chrome(service=path)
driver.get('file:///D:/selenium/book/frame.html')
#将驱动切换到 iFrame,使得 Selenium 目前可以定位
driver.switch_to.frame("frame")
#在 iFrame 内,定位登录框并输入内容
driver.find_element(By.ID,"dntest").send_keys("大牛测试")
```

代码运行后,用户名赋值成功。上面的例子是借助 iFrame id 属性来定位 iFrame 的。接下来,列举其他几种 iFrame 的定位方法。

- 通过 index 来定位,写法为 driver.switch_to.frame(0),其中 0 表示第 1 个的意思,即为第 1 个 iFrame。

- 通过 iFrame name 属性来定位,写法为 driver.switch_to.frame("daniu")。

- 通过 WebElement 对象模式来定位,即用 find_element 系列方法获取元素对象,以上 frame.html 输入用户名的代码如下:

```
driver.switch_to.frame(driver.find_element(By.ID,"dntest"))
```

注意,在上面的代码中,驱动已切换到 Frame 内部,此时只能对 Frame 内部元素进行操作,若要对 Frame 之外的元素进行操作,须将驱动切换到 Frame 之外,如在 frame.html 页面上单击"大牛测试"超链接,代码如下:

```
from selenium import webdriver
from selenium.webdriver.chrome.service import Service
path= 'D:/dr/chromedriver.exe'
from selenium.webdriver.common.by import By
path =Service('D:\\dr\\chromedriver.exe')
```

```
driver = webdriver.Chrome(service=path)
driver.get('file:///D:/selenium/book/frame.html')
#driver.switch_to.frame("frame")
driver.find_element(By.ID,"dntest").send_keys("test")
#将驱动切换回主页面
driver.switch_to.default_content()
driver.find_element(By.LINK_TEXT,"大牛测试").click()
```

5.4 上传与下载附件

在运用 Selenium 进行自动化测试的过程中，可能会遇到执行上传附件的操作。Selenium 本身是无法直接识别并操作 Windows 窗口的。在本节中，我们会列举 3 种上传附件的操作方式。

5.4.1 上传附件操作方式一

如果是 input 类型的标签且 type 为 file 类型，可以直接通过 send_keys 方法来绕过弹出框操作，直接将文件信息传递给"添加附件"按钮。下面以 upload.html 页面为例，上传附件操作如图 5.18 所示。

图 5.18

如图 5.18 所示，可以直接用 send_keys 方法操作 input 类型标签，其中在 send_keys 中填写文件路径名，代码如下：

```
driver.find_element(By.ID,"uploadattach").send_keys("D:/response.txt")
```

5.4.2　上传附件操作方式二

运用第三方工具 AutoIt 来实现上传附件。AutoIt 是免费的、运用类 Basic 语言设计开发的一款可以对 Windows 界面进行自动化模拟操作的工具。目前仅支持 Windows 操作系统，其主要有以下特点。

- Basic 语法简单易学。

- 可以模拟键盘输入和鼠标移动。

- 可以操作 Windows 窗口和任务进程。

- 可以和所有标准的 Windows 控件进行交互。

- 脚本可以被编译成单独的可执行程序，易于移植。

- 可以创建图形化用户界面。

- 支持 COM 组件。

- 支持正则表达式。

- 可以直接调用外部 DLL 库和 Windows API 方法。

AutoIt 官网提供的工具最新版本是 v6.6.x（截至本书完稿时）。该工具的安装过程很简单，此处省略。AutoIt 安装完毕后，在安装主目录中双击 "Au3Info"，会出现如图 5.19 所示的界面，这是一个悬浮窗口，可以进行元素定位，主要作用是探测 Windows 元素。在使用 AutoIt 时，用鼠标单击雷达图标 "Finder Tool"，将其拖到需要定位的 Windows 控件处即可。

如果要定位如图 5.20 所示的 Windows 窗口输入框 "File name" 和按钮 "Open"，只需要将雷达图标 "Finder Tool" 拖到相应的区域进行探测即可，如箭头所示位置。

第 5 章　Selenium 常用方法

 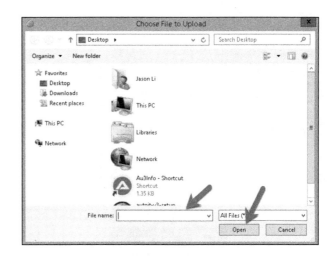

图 5.19　　　　　　　　　　　　　　　图 5.20

识别出的元素名如图 5.21 和图 5.22 所示。

- 输入框控件被识别为"Edit1"。
- 按钮控件被识别为"Button1"。

图 5.21　　　　　　　　　　　　　　　图 5.22

AutoIt 实现上传附件的步骤如下。

（1）安装 AutoIt 后，根据帮助文档 AutoIt Help File 编写测试脚本，并将其保存为

"qqattach.au3",代码如下:

```
ControlFocus("打开","","Edit")
WinWait("[CLASS:#32770]","",5)
#上传 ip.txt 文件
ControlSetText("打开","","Edit1","D:\soft\ip.txt")
Sleep(1000)
ControlClick("打开","","Button1");
```

(2)安装 AutoIt 后,打开 AutoIt v3 文件夹,单击"Compile Script to .exe (x86)",选择脚本目录、生成 exe 文件目录,并单击"Convert"按钮,如图 5.23 所示。

图 5.23

(3)打开"选择文件"对话框,并双击生成的 qqattach.au3 应用,将上传 ip.txt 文件。

5.4.3 上传附件操作方式三

第 3 种方式是通过工具 pywinauto 实现上传附件的。pywinauto 是一款对客户端系统进行自动化测试的工具类库,它是用 Python 语言编写完成的,专门处理 Windows GUI,目前仅支持 Windows 操作系统。pywinauto 的优点是可以直接用 Python 脚本调用,但需要引入相应的库。

pywinauto 直接使用命令"pip install pywinauto"进行安装,用法参见其官网上的帮助文档。下面代码的功能是利用 pywinauto 实现上传附件:

```
#coding=utf-8
#引入 Application 模块
from pywinauto.application import Application
import time
app =Application()
#定位到窗口
app = app.connect(title_re="打开",class_name="#32770")
#设置文件路径
app['打开']["EDit1"].SetEditText("D:\soft\ip.txt ")
time.sleep(2)
#单击按钮
app["打开"]["Button1"].click()
print("end")
```

5.4.4 下载附件

Selenium 支持文件或图片的下载,以 download.html 页面为例,直接在 PyCharm 中运行,如图 5.24 所示,url 为 "http://localhost:63342/pyse2023/chapter05/download.html?"。

图 5.24

对于下载功能,需要设置 ChromeOptions 属性,将 profile.default_content_settings.popups 参数设置为 "0",0 表示禁止弹出下载窗口,download.default_directory 用于设置文件存放路径,download.html 页面上的 "点击下载" 超链接的代码如下:

```
from selenium import webdriver
from selenium.webdriver.common.by import By
options = webdriver.ChromeOptions()
#图片存放路径为"/Users/tim/Desktop/selenium/book/"
prefs = {'profile.default_content_settings.popups': 0, 'download.default_
```

```
directory': '/Users/tim/Desktop/selenium/book/'}
    options.add_experimental_option('prefs', prefs)
    path = "/Users/tim/Downloads/chromedriver"
    driver = webdriver.Chrome(executable_path=path, chrome_options=options)
    driver.get('http://localhost:63342/pyse2023/chapter05/download.html?')
    driver.find_element(By.LINK_TEXT,'点击下载').click()
```

5.5 Cookie 操作

在 Web 测试过程中，常遇到 Cookie 测试，如查看不同浏览器中的 Cookie、Cookie 是否起作用等。Selenium 中提供了读取、添加、删除等 Cookie 操作方法，详细介绍如表 5.1 所示。

表 5.1

Cookie 操作方法	方 法 描 述
add_cookie(cookie_dict)	在当前会话中添加 Cookie 信息，并且参数属于字典类型数据
delete_all_cookies()	删除所有 Cookie 信息
delete_cookie(cookie_name)	删除单个名字为"cookie_name"的 Cookie 信息
get_cookie(cookie_name)	返回名为"cookie_name"的 Cookie 信息
get_cookies()	返回当前会话所有的 Cookie 信息

要更好地了解、认识 Cookie，需要清楚其工作模式。以之前登录页面为基础，对 HTML 代码进行微调，实现单击"登录"按钮后，可以添加 username 和 password 两个 Cookie。更新后的登录页面文件 logincookie.html 代码如下：

```
<!DOCTYPE html>
<html lang="en">
<head>
    <meta charset="UTF-8">
    <title>大牛测试</title>
    <script src="https://***.staticfile.org/jquery/1.10.2/jquery.min.js"></script
</head>
    <h1>大牛自动化测试</h1>
<body>
<form name="daniu">
    <label for="username">用户名：</label>
```

```html
        <input type="text" name="daniu" id="dntest" class="f-text phone-input" value="大牛" size="22">
        <label for="password">密码：</label>
        <input type="password" name="password" id="password" class ="passwd">
        <button type="submit" id ="loginbtn" onclick="document.cookie = 'username=admin;';
        document.cookie = 'password=123;';">登录</button>
        <br />
        <br>
        <a href="upload.html" target="_blank">上传资料页面</a>
        <br>
        <br>
        <label >
          <input type="radio" name="checkbox1"  value="gril" checked>女
        </label>
        <label>
          <input type="radio" name="checkbox2"  value="boy">男
        </label>
</form>
</body>
</html>
```

模拟用户登录网站前后的 Cookie 信息变化，脚本如下：

```python
#coding=utf-8
#导入 WebDriver 模块
from selenium import webdriver
from selenium.webdriver.chrome.service import Service
import time
#这是在 macOS 上执行的基于 Chrome 浏览器的测试，请根据实际情况修改如下的 driver 地址
chrome_driver_server = Service("./chromedriver")
driver = webdriver.Chrome(service=chrome_driver_server)
driver.maximize_window()
#打开登录主页
driver.get('http://localhost:63342/selenium4.0-automation/logincookie.html')
print("before login:")
#打印全部 Cookie
for cookie_detail in driver.get_cookies():
    print(cookie_detail)
#等待 30s，方便手动干预输入用户名 admin、密码 123 和单击"登录"按钮
time.sleep(30)
```

```
print("after login:")
for cookie_detail in driver.get_cookies():
    print(cookie_detail)
driver.quit()
```

注意，登录后的 Cookie 信息中多了 username 和 password 这两个 Cookie，具体如图 5.25 所示。

```
before login:
{'domain': 'localhost', 'expiry': 1711287581, 'httpOnly': True, 'name': 'Pycharm-384973f8', 'path': '/', 'sameSite': 'Strict', 'secure': False, 'value': '89098811-dfec-4f63-8d48-464276971d7f'}
after login:
{'domain': 'localhost', 'expiry': 1711287581, 'httpOnly': True, 'name': 'Pycharm-384973f8', 'path': '/', 'sameSite': 'Strict', 'secure': False, 'value': '89098811-dfec-4f63-8d48-464276971d7f'}
{'domain': 'localhost', 'httpOnly': False, 'name': 'password', 'path': '/selenium4.0-automation', 'sameSite': 'Lax', 'secure': False, 'value': '123'}
{'domain': 'localhost', 'httpOnly': False, 'name': 'username', 'path': '/selenium4.0-automation', 'sameSite': 'Lax', 'secure': False, 'value': 'admin'}
```

图 5.25

通过上面获取到的 Cookie，可以实现自动登录网站的目的。把 Cookie 保存到 list 中，用 add_cookie 方法将其添加到网站中，代码如下：

```
#coding=utf-8
from selenium import webdriver
from selenium.webdriver.chrome.service import Service
import time
#这是在 macOS 上执行的基于 Chrome 浏览器的测试，请根据实际情况修改如下的 driver 地址
chrome_driver_server = Service("./chromedriver")
driver = webdriver.Chrome(service=chrome_driver_server)
driver.maximize_window()
driver.get('http://localhost:63342/selenium4.0-automation/logincookie.html')  #打开登录页面
#根据之前打印的登录网站之后的 Cookie 信息，手动添加 Cookie，使用这种方式可以跳过验证码
#只需要将下面 Cookie 的 value 值设置为之前登录后获取的相应的值即可
coo =[{'domain': 'localhost', 'expiry': 1711287581, 'httpOnly': True, 'name': 'Pycharm-384973f8', 'path': '/', 'sameSite': 'Strict', 'secure': False, 'value': '89098811-dfec-4f63-8d48-464276971d7f'},
      {'domain': 'localhost', 'httpOnly': False, 'name': 'username', 'path': '/selenium4.0-automation', 'sameSite': 'Lax', 'secure': False, 'value': 'admin'},
      {'domain': 'localhost', 'httpOnly': False, 'name': 'password', 'path': '/selenium4.0-automation', 'sameSite': 'Lax', 'secure': False, 'value': '123'}
     ]
for cookie in coo:
    driver.add_cookie(cookie)
time.sleep(3)
driver.get("http://localhost:63342/selenium4.0-automation/logincookie.html")
```

5.6 驱动管理模式

Selenium 4 支持驱动管理模式,使用 WebDriver Manager 模式进行驱动管理时首先需要下载 WebDriver Manager 模块,命令为"pip install webdriver-manager"。

模块下载完成后,可以使用如下示例代码来实现对驱动的使用。其中 install 方法完成了对对应浏览器驱动的下载和管理。

```
from selenium import webdriver
from selenium.webdriver.chrome.service import Service
from webdriver_manager.chrome import ChromeDriverManager
driver = webdriver.Chrome(service=Service
(ChromeDriverManager().install()))
driver.get("file:///Users/tim/Desktop/selenium/book/login.html")
```

以上测试代码执行后,日志中显示 WebDriver Manager 获取了本地浏览器的版本,再联网下载相应的驱动,这种方式的特色是可以自动下载驱动。

5.7 颜色验证

UI 元素相关的颜色验证有时也属于项目需求,如何验证颜色呢?可以分为如下两个步骤。

1. 获取元素颜色

可以通过 Inspect 方法获取与元素相关的颜色数值。例如,获取 hover.html 系统的背景色,如图 5.26 所示,元素背景色的 Hex Color 值为"#7fffd4"。

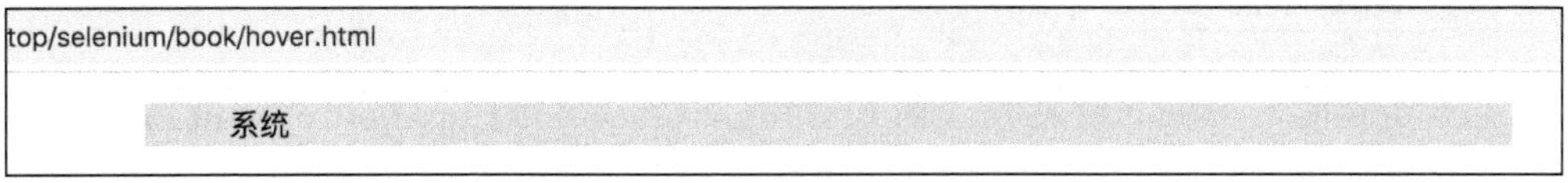

图 5.26

2. 颜色验证代码演示

获取元素"系统"的背景色的代码如下，运行代码后会输出字符"#7fffd4"，证明颜色验证通过：

```python
from selenium import webdriver
from selenium.webdriver.support.color import Color
from selenium.webdriver.chrome.service import Service
from selenium.webdriver.common.by import By
driver = webdriver.Chrome()
driver.get("file:///Users/tim/Desktop/selenium/book/hover.html")
#获取元素的背景色#7fffd4
bg_color = Color.from_string(driver.find_element(By.XPATH,' //*[@id="testdn"]').value_of_css_property('background-color'))
print(bg_color.hex)
```

5.8　3 种等待模式

Selenium 提供了 3 种不同类别的等待模式，可在识别元素或等待页面加载时使用它们。

5.8.1　强制等待模式

强制等待模式使用了 Python 官方自带的 time 模块，写法为 time.sleep(n)，表示强制等待 n 秒。这种模式多用于调试脚本，简单易用。缺点也比较明显，如果使用这种等待模式，那么可能会出现等待时间到了，元素定位等操作还没完成的情况，也可能会出现操作已经完成，但是等待时间未耗尽的情况。

5.8.2　隐式等待模式

在元素定位时，Selenium 提供了隐式等待模式，写法为 implicitly_wait(T)，表明在 T 时间内，只有完成了整个页面的加载，才执行下一步操作。在设定的时间内没有完成页面的加载，会在时间耗尽后，执行下一步操作。隐式等待语句在 driver 对象的生命周期内都是有效的，只需要写明一次即可。以 login.html 页面为例，代码如下：

```
from selenium import webdriver
from selenium.webdriver.support.color import Color
from selenium.webdriver.chrome.service import Service
from selenium.webdriver.common.by import By
import time
driver = webdriver.Chrome()
driver.get("file:///D:/selenium/book/login.html")
#表示设置的隐式等待时间为10s, 即最长等待时间为10s
driver.implicitly_wait(10)
driver.find_element(By.ID,'dntest').send_keys("大牛测试")
```

隐式等待模式也有一些缺点，比如这种模式需要等待页面加载完成后才能执行下一步操作，所以从这一点看，脚本的执行不是很高效。

5.8.3 显式等待模式

显式等待模式主要会将页面某个元素作为判断条件。它的实现需要用到 WebDriverWait 类，该类有 3 个参数，如 WebDriverWait(driver,10,1)表示最长等待时间为 10s，每隔 1s 检查 1 次元素，直至元素被定位。在实际使用中，该类需要结合 expected_conditions 模块，常用的 expected_conditions 方法如下。

- title_is，判断当前页面的 title 是否等于某个值。
- presence_of_element_located，判断某个 locator 元素是否被加到 DOM 树里，可应用于可见或隐藏元素。
- element_selection_state_to_be，检查元素是否被选中。
- visibility_of，检查元素是否能够被看到。
- visibility_of_element_located，判断某个 locator 元素是否可见。
- element_to_be_clickable，判断某个 locator 元素是否可点击。
- element_to_be_selected，判断元素是否被选中，一般用在下拉列表中。
- element_located_to_be_selected，根据定位语句判断元素是否被选中。

例如，实现 login.html 页面时，判断登录框是否可见，visibility_of_element_located 使用的是 locate 变量，locate 须传入元组类型参数，包含定位方法与属性值，代码如下：

```
from selenium.webdriver.chrome.service import Service
```

```
from selenium.webdriver.common.by import By
from webdriver_manager.chrome import ChromeDriverManager
from selenium import webdriver
from selenium.webdriver.support.ui import WebDriverWait
from selenium.webdriver.support import expected_conditions as ex
path = Service("D:\\dr\\chromedriver.exe")
driver = webdriver.Chrome(service=path)
#driver.implicitly_wait(10)
driver.get("file:///D:/selenium/book/login.html")
locate = (By.ID,"dntest")
WebDriverWait(driver,10,1).until(ex.visibility_of_element_located(locate))
```

5.9 多窗口切换

在自动化测试过程中，经常会遇到从一个窗口切换到一个新窗口的场景，即使打开一个新窗口，但 driver 还停留在原窗口，这时仍无法对新窗口控件进行操作，Selenium 提供了句柄方法，可以很好地解决这类问题。一个句柄就是一个字符串，可以唯一地标识当前窗口，需要注意的是，每次运行时句柄值都会发生变化。以在 login.html 页面上单击"上传资料页面"，在 upload 页面上打开上传附件窗口为例，代码如下：

```
from selenium import webdriver
from selenium.webdriver.chrome.service import Service
from selenium.webdriver.common.by import By
path= Service('D:/dr/chromedriver.exe')
driver = webdriver.Chrome(service=path)
driver.get('file:///D:/selenium/book/login.html#')
#打印当前窗口句柄
print(driver.current_window_handle)
driver.find_element(By.PARTIAL_LINK_TEXT,"上传资料").click()
#打印所有窗口句柄
print(driver.window_handles)
#切换到新页面
driver.switch_to.window(driver.window_handles[-1])
#打开上传附件窗口
driver.find_element(By.ID,"AttachFrame").click()
```

运行结果如图 5.27 所示，"CDwindow-3100927A74CF0AD5F2E126B0849C4E2F"就是此次运行 login.html 页面的句柄，使用 window_handles 打印所有窗口句柄，window_handles 方法的输出为列表类型，找到新窗口句柄之后用 switch_to.window 方法切换到新窗口：

```
/usr/local/bin/python3 /Users/tim/Documents/pyse2023/chapter05/daniu_5.9.1.py
CDwindow-3100927A74CF0AD5F2E126B0849C4E2F
['CDwindow-3100927A74CF0AD5F2E126B0849C4E2F', 'CDwindow-3CD90EFC9BC70DC6DFEBE35DF2728C53']

Process finished with exit code 0
```

图 5.27

5.10 弹框操作

HTML 中有 3 种弹框，分别是 alert 警示对话框、confirm 确认对话框和 prompt 提示框，以 a1.html 页面为例，alert 警示对话框如图 5.28 所示。

图 5.28

（1）alert 警示对话框代码如下：

```
from selenium import webdriver
from selenium.webdriver.chrome.service import Service
from selenium.webdriver.common.by import By
path = Service("D:\\dr\\chromedriver.exe")
driver = webdriver.Chrome(service=path)
driver.maximize_window()
driver.get("file:///D:/selenium/book/a1.html")
#打开alert警示对话框
driver.find_element(By.ID,"alert").click()
#切换到alert警示对话框
```

```
a1 = driver.switch_to.alert
#执行接受操作,即消除弹框
a1.accept()
```

(2) confirm 确认对话框如图 5.29 所示。

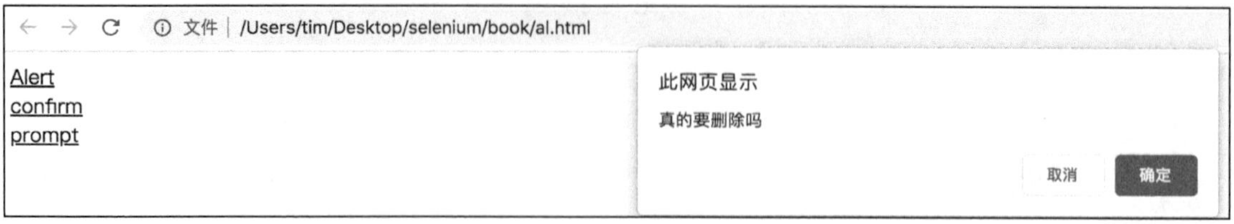

图 5.29

switch_to.alert 方法切换 driver 至弹框页面,accept 方法执行接受操作,dismiss 方法执行取消操作,代码如下:

```
from selenium import webdriver
from selenium.webdriver.chrome.service import Service
from selenium.webdriver.common.by import By
path = Service("D:\\dr\\chromedriver.exe")
driver = webdriver.Chrome(service=path)
driver.maximize_window()
driver.get("file:///D:/selenium/book/a1.html")
driver.find_element(By.ID,"confirm").click()
a1 = driver.switch_to.alert
#执行接受操作,消除弹框
a1.accept()
#执行取消操作
#a1.dismiss()
```

(3) prompt 提示框如图 5.30 所示。

图 5.30

使用 send_keys 方法可往提示框中输入内容,text 方法用于输出文字"请输入名字",accept 方法用于单击"确定"按钮,dismiss 方法用于单击"取消"按钮,代码如下:

```
from selenium import webdriver
from selenium.webdriver.chrome.service import Service
from selenium.webdriver.common.by import By
path = Service("D:\\dr\\chromedriver.exe")
driver = webdriver.Chrome(service=path)
driver.maximize_window()
driver.get("file:///D:/selenium/book/al.html")
#打开prompt提示框
driver.find_element(By.ID,"prompt").click()
#切换到prompt提示框
al = driver.switch_to.alert
print(al.text)
#al.dismiss()
#往提示框中输入内容
al.send_keys("dntest")
#执行接受操作,消除弹框
al.accept()
```

5.11 ChromeOptions

ChromeOptions 是一个配置 Chrome 启动时属性的类。通过这个类,可以为 Chrome 配置如下参数,Selenium 4 通过 add_argument 方法增加参数,常用参数如下。

- "start-fullscreen",表示全屏。
- "headless",表示无界面运行。
- "window-size=400,600",表示设置窗口尺寸为 400×600。
- "kiosk",表示窗口无地址栏。

以 login.html 页面为例实现全屏操作,需要注意的是,在使用 Chrome 时要增加 options 参数,代码如下:

```
from selenium import webdriver
from selenium.webdriver.chrome.service import Service
path = Service("D:\\dr\\chromedriver.exe")
options = webdriver.ChromeOptions()
#全屏
options.add_argument("start-fullscreen")
#在Chrome方法中增加options参数
```

```
driver = webdriver.Chrome(service=path,options=options)
#打开login.html页面
driver.get("file:///D:/selenium/book/login.html")
```

5.12 滑块操作

在登录或者注册页面时，经常采用滑块作为安全验证机。本节会通过一个具体案例来学习自动化处理滑块的思路，以 sliding_block.html 页面为例，如图 5.31 所示

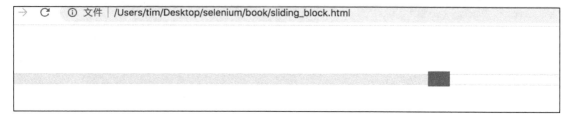

图 5.31

当处于页面初始状态时，滑块（小长方形模块）处在滑槽正中，并且值显示为 50。在操作滑块的过程中，需要确定起始点元素与滑块轨道坐标，以下代码实现了将页面上的滑块从正中位置滑动到最右边，并且值显示为 100.00：

```
from selenium import webdriver
from selenium.webdriver.chrome.service import Service
from selenium.webdriver import ActionChains
from selenium.webdriver.common.by import By
import time
#这是在macOS上执行的基于Chrome浏览器的测试，请根据实际情况修改如下的driver地址
chrome_driver_server = Service("/Users/xxx/Downloads/chromedriver")
driver  = webdriver.Chrome(service=chrome_driver_server)
driver.maximize_window()
driver.get("file:///Users/xxx/PycharmProjects/selenium4.0-automation/Chapter7/selenium_sliding_block.html")
source_element = driver.find_element(By.ID,'sliding1')
target_element = driver.find_element(By.ID,'range1')
target_element_X_Offset = target_element.location.get("x")
target_element_Y_Offset = target_element.location.get("y")
```

```
#print(target_element_X_Offset)
#print(target_element_Y_Offset)
#print(source_element.location.get("x"))
#print(source_element.location.get("y"))
time.sleep(3)
ActionChains(driver).drag_and_drop_by_offset(source_element,700,0).perform()
driver.quit()
```

以上代码主要用到了 drag_and_drop_by_offset 方法,执行结果如图 5.32 所示。

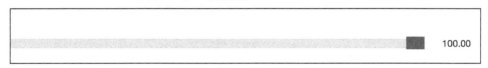

图 5.32

5.13 元素截图

在自动化测试中,处理验证码功能时,常采用图像识别的方式,这种方式需要先对元素进行截图,而截图功能主要有如下两种方式。

(1)使用元素自带的 screenshot 方法,以 login.html 页面登录功能为例,执行 screenshot 方法后当前目录下生成了 login.png 图片,代码如下:

```
from selenium import webdriver
from selenium.webdriver.chrome.service import Service
from selenium.webdriver.common.by import By
driver = webdriver.Chrome()
driver.get("file:///Users/tim/Desktop/selenium/book/login.html")
driver.save_screenshot("verification_code.png")
imglogin=driver.find_element(By.ID,"img")
imglogin.screenshot("login.png")
```

(2)先截整张图片,再进行裁切,这时需要安装 pillow 库,安装 pillow 库的命令为"pip install pillow",同样以 login.html 页面登录功能为例,代码如下:

```
from selenium import webdriver
from selenium.webdriver.common.by import By
```

```python
from PIL import Image
driver = webdriver.Chrome()
driver.get("file:///Users/tim/Desktop/selenium/book/login.html")
time.sleep(2)
driver.save_screenshot("code.png")
imgcode=driver.find_element(By.ID,"loginbtn")
#登录位置left值
left= imgcode.location['x']
#登录位置top值
top= imgcode.location['y']
#加上宽度
right = left+imgcode.size['width']
#加上高度
bottom = top+imgcode.size['height']
im = Image.open("vcode.png")
im = im.crop((left,top,right,bottom))
im.save('login.png')
```

5.14 JavaScript 操作页面元素

WebDriver 对部分浏览器上的控件并不直接提供支持，如浏览器右侧滚动条、副文本等，而是借助 JavaScript 间接操作。WebDriver 提供了 execute_script 和 execute_async_scrip 两种方法来执行 JavaScript 代码，其区别如下。

（1）execute_script 方法是同步执行方法且执行时间较短。WebDriver 会等待同步执行的结果，然后执行后续代码。

（2）execute_async_script 方法是异步执行方法且执行时间较长。WebDriver 不会等待异步执行代码的结果，而是直接执行后续的代码。

以 login.html 页面为例，打开谷歌浏览器的开发者工具进入 Console 工具页面，用 document.getElementById('').value 方法对用户名输入框进行操作。具体做法：在 Console 页面键入代码"document.getElementById('dntest').value='测试'"，然后按下回车键，就可以实现如图 5.33 所示的效果。相关代码如下：

```
from selenium import webdriver
```

```
from selenium.webdriver.chrome.service import Service
#driver 地址
chrome_driver_server = Service("/Users/xxx/Downloads/chromedriver")
driver = webdriver.Chrome(service=chrome_driver_server)
driver.maximize_window()
driver.get("file:///D:/selenium/book/login.html")
js = "document.getElementById('dntest').value = '测试'"
driver.execute_script(js)
#driver.quit()
```

图 5.33

使用 JavaScript 代码来执行对浏览器滚动条的操作，以 scroll.html 页面为例，代码如下：

```
#coding=utf-8
from selenium import webdriver
from selenium.webdriver.chrome.service import Service
#driver 地址
path =Service('D:\\dr\\chromedriver')
driver = webdriver.Chrome(service=path)
driver.maximize_window()
driver.get("file:///D:/selenium/book/scroll.html")
#设置浏览器窗口大小，目的是让滚动条显示出来
driver.set_window_size(600,400)
js = "window.scrollTo(100,300)"
driver.execute_script(js)
```

运行代码后的效果如图 5.34 所示。

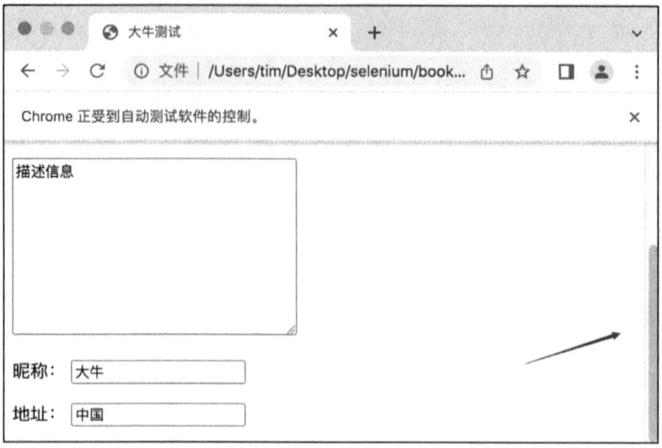

图 5.34

注意：

（1）使滚动条滑到底部，可使用以下的 JavaScript 代码：

`"window.scrollTo(0,document.body.scrollHeight)"`

（2）直接互换横向滚动条的 x、y 坐标值，如将横向滚动条滑到最右侧：

`window.scrollTo(document.body.scrollHeight,0)`

（3）自定义属性，在自动化测试过程中，有时需要定义一个临时属性，用 JavaScript 代码可以很好地实现该功能。

如给 login.html 页面用户名输入框增加一个属性名 daniu，属性值是 daniutest，对应的 JavaScript 代码如下：

`document.getElementById('dntest').setAttribute("daniu","daniutest");`

运行代码后用户名标签增加了属性 daniu="daniutest"，如图 5.35 所示，代码如下：

```
from selenium import webdriver
from selenium.webdriver.chrome.service import Service
from selenium.webdriver.common.by import By
#driver 地址
path =Service('D:\\dr\\chromedriver')
driver  = webdriver.Chrome(service=path)
driver.maximize_window()
driver.get("file:///D:/selenium/book/login.html")
js =  "document.getElementById('dntest').
setAttribute('daniu','daniutest');"
```

```
driver.execute_script(js)
#输出 daniu 新属性
print(driver.find_element(By.ID,"dntest").get_attribute("daniu"))
```

```
<!DOCTYPE html>
<html lang="en">
▶ <head>...</head>
▼ <body style> == $0
    <h1>大牛自动化测试</h1>
    <form name="daniu">
        <label for="username">用户名: </label>
        <input type="text" name="daniu" id="dntest" class="f-text phone-input" value="大牛" size="22" daniu="daniutest">
        <label for="password">密码: </label>
        <input type="password" name="password" id="password" class="passwd">
        <button type="submit" id="loginbtn">登录</button>
        <br>
```

图 5.35

5.15 jQuery 操作页面元素

jQuery 是 JavaScript 的一个类库，在 JavaScript 的基础上进行了深度封装。与 JavaScript 相比，jQuery 可以用更少的代码实现同样的功能。它是元素定位的另一个强有力的补充。就像打通了元素定位的任督二脉一样，会让用户在使用 Selenium 的时候更加得心应手。jQuery 选择器的基本语法结构如图 5.36 所示。

选择器	实例	选取
jQuery 选择器		
*	$("*")	所有元素
#id	$("#lastname")	id="lastname" 的元素
.class	$(".intro")	所有 class="intro" 的元素
element	$("p")	所有 <p> 元素
.class.class	$(".intro.demo")	所有 class="intro" 且 class="demo" 的元素
:first	$("p:first")	第一个 <p> 元素
:last	$("p:last")	最后一个 <p> 元素
:even	$("tr:even")	所有偶数 <tr> 元素
:odd	$("tr:odd")	所有奇数 <tr> 元素
:eq(index)	$("ul li:eq(3)")	列表中的第四个元素（index 从 0 开始）
:gt(no)	$("ul li:gt(3)")	列出 index 大于 3 的元素
:lt(no)	$("ul li:lt(3)")	列出 index 小于 3 的元素
:not(selector)	$("input:not(:empty)")	所有不为空的 input 元素

图 5.36

同上面的 JavaScript 的例子，在 login.html 页面用户名输入框中输入文本"测试"，在 Console 页面执行"$('#dntest').val('测试')"后，用户名输入框中输入了"测试"。执行效果如图 5.37 所示。

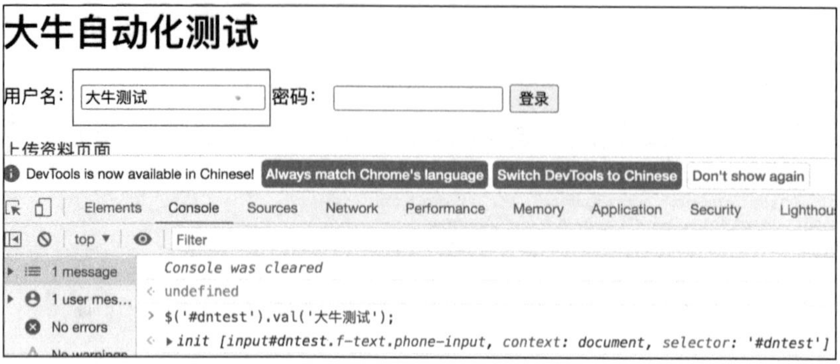

图 5.37

代码运行方式同 JavaScript，代码如下：

```
#coding=utf-8
from selenium import webdriver
from selenium.webdriver.chrome.service import Service
from selenium.webdriver.common.by import By
path =Service('D:\\dr\\chromedriver')
#driver 地址
driver = webdriver.Chrome(service=path)
driver.maximize_window()
driver.get("file:///D:/selenium/book/login.html")
jq = "$('#dntest').val('测试')"
driver.execute_script(jq)
```

5.16 innerText 与 innerHTML

innerText 属性将文本内容设置或返回为指定节点及其所有子节点的纯文本，如 inner.html 页面对标签<p>操作，输出字符"大牛测试"，如图 5.38 所示。

innerHTML 属性将获取和设置元素中的纯文本或 HTML 内容。与 innerText 不同，InnerHTML 允许使用 HTML 格式的文本，并且不会自动对文本进行编码和解码，如

inner.html 页面对标签<p>操作，输出"大牛测试"，如图 5.38 所示。

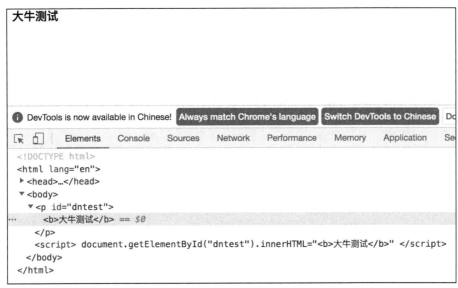

图 5.38

上例中也可以用 get_attribute 方法获取 innerText 与 innerHTML 的值，代码如下：

```
from selenium import webdriver
from selenium.webdriver.chrome.service import Service
#driver 地址
from selenium.webdriver.common.by import By
path =Service('D:\\dr\\chromedriver')
driver  = webdriver.Chrome(service=path)
driver.maximize_window()
driver.get("file:///D:/selenium/book/inner.html")
#输出 innerText 值"大牛测试"
print(driver.find_element(By.ID,"dntest").get_attribute("innerText"))
#输出 innerHTML 值"<b>大牛测试</b>"
#print(driver.find_element(By.ID,"dntest").get_attribute("innerHTML"))
```

5.17　通过源码理解 By.ID

前面学习了 Selenium 八大定位，即通过 By.×× 实现定位，我们进行下面的测试，直接输出 By.ID，代码如下：

```
#coding=utf-8
from selenium import webdriver
from selenium.webdriver.chrome.service import Service
from selenium.webdriver.common.by import By
from selenium.webdriver import ActionChains
path =Service('D:\\dr\\chromedriver')
driver = webdriver.Chrome(service=path)
print(By.ID)
```

运行代码后输出的结果如图 5.39 所示，输出值为 id。

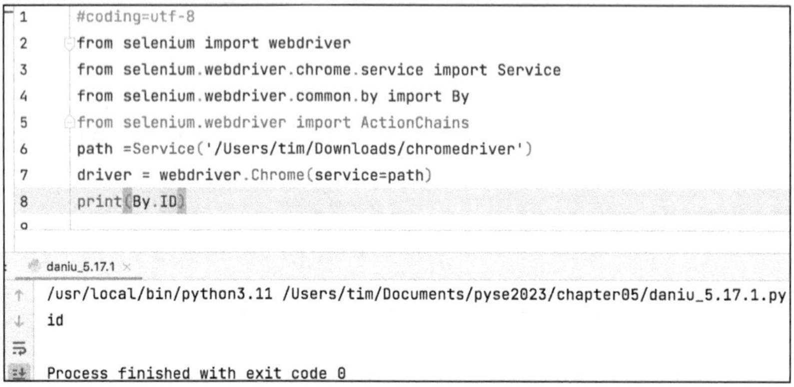

图 5.39

要理解为什么输出值为 id，需要查看源码，按住 Ctrl 键移动鼠标至源码 By 并单击，代码如下：

```
class By:
    """
    Set of supported locator strategies.
    """
    ID = "id"
    XPATH = "xpath"
    LINK_TEXT = "link text"
    PARTIAL_LINK_TEXT = "partial link text"
    NAME = "name"
    TAG_NAME = "tag name"
    CLASS_NAME = "class name"
    CSS_SELECTOR = "css selector"
```

从上面的代码可知 By 为一个类，含有 8 个变量，其中 ID 变量为"id"，故 By.ID 输出的 ID 值即"id"。ID 的类型为字符串，以 login.html 页面用户名为例，尝试用"id"替换 By.ID，代码如下：

```
#coding=utf-8
from selenium import webdriver
from selenium.webdriver.chrome.service import Service
from selenium.webdriver.common.by import By
from selenium.webdriver import ActionChains
path =Service('/Users/tim/Downloads/chromedriver')
driver = webdriver.Chrome(service=path)
driver.get("file:///Users/tim/Desktop/selenium/book/login.html")
driver.find_element("id","dntest").send_keys("测试")
```

运行代码后用户名成功变为"大牛测试"，如图 5.40 所示。

图 5.40

通过本章的学习，读者可以掌握 Selenium 的常用方法，包括熟悉每种方法的使用场景或者前提条件。

第三篇

项 目 篇

归根结底，学习知识的目的就是为了应用，在实战的过程中可以学到更深入的知识。结合实际项目的学习能够更好地理解理论知识，深化对理论知识的认识。如同编程一样，刚开始项目很简单、很小，慢慢地随着学习的深入，我们可以迭代项目代码，扩充、完善项目功能。在不断地试错、迭代、完善项目的同时，项目的轮廓也会越来越清晰。

初学者做项目不要贪多、贪大。先分析需求点，然后一点一点思考怎么实现项目，解析、分割需求时，最好做到功能模块的独立性，这样可以降低程序的耦合程度。对于自动化测试来说，低耦合可以增强自动化框架的可扩展性。

从简单到复杂，分析测试需求，实现项目需求，以及最后优化项目需求，这是一个逐步深入的过程。本篇对应的章节如下。

第 6 章　项目实战

第 7 章　项目重构与代码优化

第 8 章　数据驱动测试

第 9 章　Page Object 设计模式

第 10 章　pytest 框架实战

第 11 章　行为驱动测试

第 6 章 项目实战

本章通过项目实战的方式帮助读者树立对自动化测试的整体认识。通过项目实战可以更快地将基础知识串联起来，帮助初学者快速成长。

6.1 项目需求分析汇总

需求分析是分析项目要覆盖的业务场景，以及要实现自动化的功能点，是开始实施自动化测试的必要步骤。在做自动化项目时，需要进行前期调研，根据调研结果来确定项目相关的属性，如适用自动化测试的范围、测试方法、测试策略等。根据项目的不同，需求分析的侧重点也会不一样。

6.1.1 制订项目计划

如何有效地管理上面提到的条目呢？答案是制订一个有效的项目计划。项目计划至关重要，它的好坏直接决定项目能否成功，做计划就是为了预防异常风险项等。

一定要有项目范围，把范围定义得清晰至关重要。假如项目范围不清晰，则会导致验证标准无法确定。如某次自动化测试的项目范围仅是测试某网站的登录功能，这时便不需要考虑网站的其他功能。

项目目标设定，需要综合评估项目本身，如范围，涉及的业务类型、大小、风险评估等。综合评估之后，才可设定一个较合理的目标。

项目目标设定完成后，需要规划一些活动来支持和完成既定目标。规划活动时需要围绕项目目标，否则就没有意义，应避免规划出无效的活动。为了项目目标，需要将无效或者低效的活动从项目规划中剔除。

为项目做计划，没有标准答案。同一个项目，不同的人最终做出来的项目计划肯定不同。读者应该在不同的、多种多样的项目中练习做项目计划。

项目执行时需要遵循项目计划，如果计划没有一项一项地落实，那么项目计划就无法发挥实质性的作用。项目计划成熟的标志是，它具有独立性和非主观性，即使让非项目计划制订者去执行，也可以顺利执行。

为了帮助大家从零到一地落地自动化项目，接下来将以大牛测试后台项目为例，一步步实施自动化测试。首先搭建好测试环境，步骤如下。

- 在 Windows 系统中安装好 JDK 11。
- 在配套资源包中，下载文件 dntest.jar。
- 执行命令"java -jar dntest.jar"。
- 打开网站"http://localhost/login"。
- 输入用户名"admin"，密码"admin123"。

在本章的项目实战中，项目计划如表 6.1 所示，由于篇幅限制，对该项目计划进行了精简，只列出了关键项，没有列出时间、人员等信息。在实际的项目中，应根据项目的实际情况来创建项目计划。项目计划没有固定的格式和内容要求。

表 6.1

项 目 简 介
用自动化方式实现后台管理系统中对系统模块进行常规性功能测试
项目启动前置条件
1）后台管理系统工作正常
2）自动化测试环境准备完毕（Python、Selenium 和 PyCharm 安装完毕）
项目覆盖场景
场景的确定需要根据性能需求分析得出。这样的过程需要多方人员参与，如开发人员、测试人员、产品经理、项目经理等
1）大牛测试系统登录
2）新增岗位信息
3）新增部门信息
4）订单信息页面
5）新增用户信息

6.1.2 编写测试用例

对于自动化测试，也需要编写测试用例，这是规范和追踪测试活动的必不可少的环节。切忌在没有测试用例的情况下直接编写脚本，或者直接将手工测试的测试用例作为自动化测试用例。在编写自动化测试用例时，要做到以下两点。

- 简明扼要，便于理解和易于执行。
- 设计的测试用例需要全面覆盖场景，如边界值、正反例等。

请参考如下测试用例实例，如表 6.2 所示。

表 6.2

ID	模 块 名	覆盖功能点	前 置 条 件	测试步骤	预 期 结 果
testcase_01	系统登录	登录功能	登录功能正常	（1）打开系统登录页面。 （2）在用户名和密码输入框中分别输入用户名与密码。 （3）使用验证码识别技术识别验证码。 （4）输入第（3）步中识别出来的验证码。 （5）单击"登录"按钮	（1）登录页面能正常打开。 （2）用户名、密码信息能正常输入。 （3）验证码能被正确识别。 （4）识别出来的验证码能被正确输入。 （5）用户能正常登录系统

续表

ID	模块名	覆盖功能点	前置条件	测试步骤	预期结果
testcase_02	岗位管理	岗位信息管理功能	管理员能正常登录系统	（1）打开岗位管理页面。 （2）单击"新增"按钮。 （3）在添加岗位页面，输入"岗位名称""岗位编码""显示顺序""备注"等。 （4）单击"确定"按钮	（1）岗位管理页面能正常打开。 （2）添加岗位页面能正常打开。 （3）测试步骤（3）中的这些字段都能输入相关字段内容。 （4）能够成功添加岗位
testcase_03	部门管理	部门信息管理功能	管理员能正常登录系统	（1）打开部门管理页面。 （2）单击"新增"按钮。 （3）在添加部门页面，输入"上级部门""部门名称""显示顺序""负责人""联系电话""邮箱""部门状态"（正常或停用）等。 （4）单击"确定"按钮	（1）部门管理页面能正常打开。 （2）添加部门页面能正常打开。 （3）测试步骤（3）中的这些字段都能输入相关字段内容。 （4）能够成功添加部门
testcase_04	角色管理	角色信息管理功能	管理员能正常登录系统	（1）打开角色管理页面。 （2）单击"新增"按钮。 （3）在添加角色页面，输入"角色名称""权限字符""显示顺序""状态""备注""菜单权限"等。 （4）单击"确定"按钮	（1）角色管理页面能正常打开。 （2）添加角色页面能正常打开。 （3）测试步骤（3）中的这些字段都能输入相关字段内容。 （4）能够成功添加角色
testcase_05	用户管理	用户信息管理功能	管理员能正常登录系统	（1）打开用户管理页面。 （2）单击"新增"按钮。 （3）在添加用户页面，输入"用户名称""手机号码""登录账户""用户性别""岗位""角色""归属部门""邮箱""登录密码""用户状态"等。 （4）单击"确定"按钮	（1）用户管理页面能正常打开。 （2）添加用户页面能正常打开。 （3）测试步骤（3）中的这些字段都能输入相关字段内容，其中岗位、角色和归属部门可以选择 testcase_02/03/04 新建的记录。 （4）能够成功添加用户

我们以大牛测试系统的登录和新增岗位信息为例来展开介绍怎么做需求分析。需求分析内容主要是对覆盖的业务场景和对每个页面上的关键元素的分析。主流程如图 6.1 所示。

第 6 章　项目实战

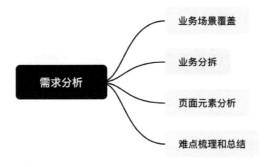

图 6.1

以上是对项目需求的简单分析，通过流程图能更清晰地勾勒出项目的需求点，以便测试参与者有整体认识。

6.2　业务场景的覆盖与分拆

针对前面介绍的需求分析步骤，业务场景覆盖与分拆是项目前提，接下来会进行详细说明。

测试场景"testcase_01"，被测试功能涉及的页面描述如下。

大牛测试系统登录页面如图 6.2 所示，页面 URL 为 http://localhost/login。

图 6.2

输入用户名、密码和验证码后单击"登录"按钮，成功登录系统，登录成功如图 6.3 所示。

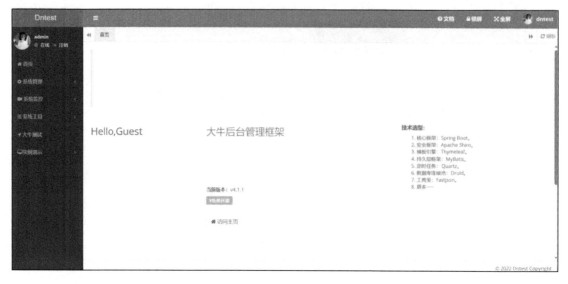

图 6.3

测试场景"testcase_02"，被测试功能涉及的页面描述如下。在系统首页单击"系统管理->岗位管理"，岗位管理页面如图 6.4 所示。

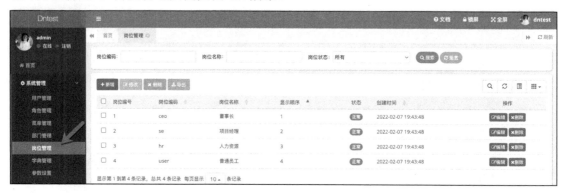

图 6.4

单击"新增"按钮，打开添加岗位页面，如图 6.5 所示。

图 6.5

以上是对项目中涉及的页面的粗略统计和认识，共涉及 4 个页面，具体如下。

- 系统登录页面

- 系统首页

- 岗位管理页面

- 添加岗位页面

6.2.1　逐个分析页面元素

在系统登录页面，如图 6.2 所示，对项目中涉及的元素逐一进行分析。需要结合元素自身的实际情况来选择元素定位方法，而不一定要用某一特定的定位方法。元素定位方法有一定的优先级，如优先选择 id 定位方法，须注意 id 属性值在页面上必须是唯一的，接下来才是 name、class、CSS、XPath 等定位方法。

1. 用户名输入框

用户名具体元素如图 6.6 中的灰底部分所示,可以通过 name 定位方法来定位元素,值为"username"。

图 6.6

2. 密码输入框

密码具体元素如图 6.7 中的灰底部分所示,可以通过 name 定位方法来定位元素,值为"password"。

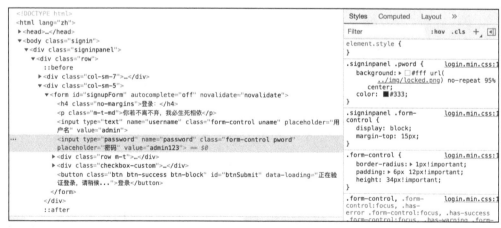

图 6.7

3. 验证码图片

验证码具体元素如图 6.8 中的灰底部分所示，可以通过 XPath 定位方法来定位元素，值为"//*[@id="signupForm"]/div[1]/div[2]/a/img"，作用是获取验证码截图，以便用验证码识别技术进行识别。

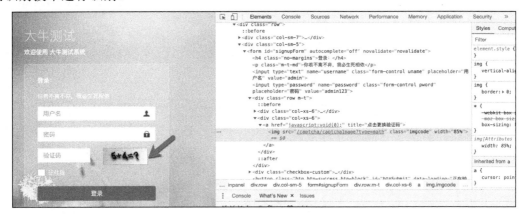

图 6.8

4. 验证码输入框

验证码是一个输入框元素，用于输入前一步验证码图片的识别结果，元素具体细节如图 6.9 所示。可以通过 name 定位方法来定位元素，值为"validateCode"。

图 6.9

5."登录"按钮

"登录"按钮元素用于提交登录信息，元素细节如图 6.10 所示。可以通过 id 定位方法来定位元素，值为"btnSubmit"。

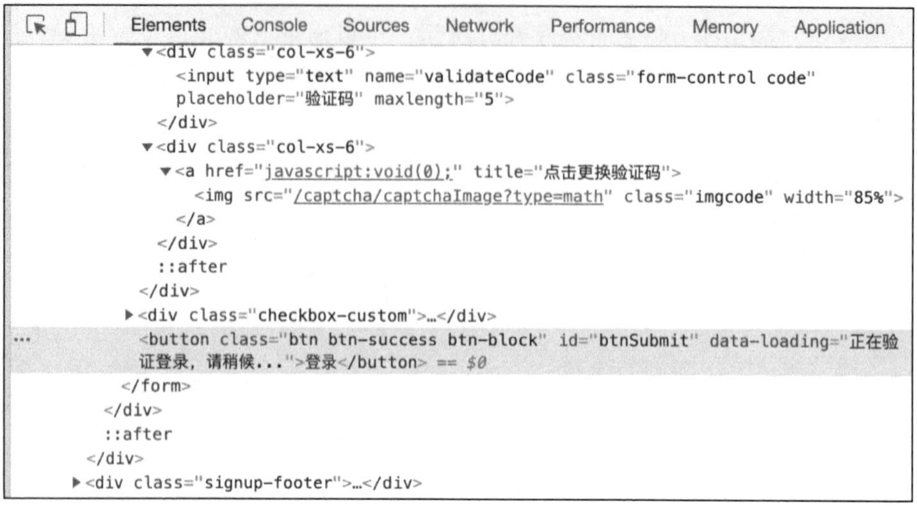

图 6.10

系统首页，如图 6.3 所示。

6. 系统管理 Span 元素

"系统管理"是左侧导航栏一级菜单项，是一个 HTML Span 元素，元素细节如图 6.11 所示。可以通过 XPath 定位方法来定位元素，值为"//*[@id="side-menu"]/li[3]/a/span[1]"。

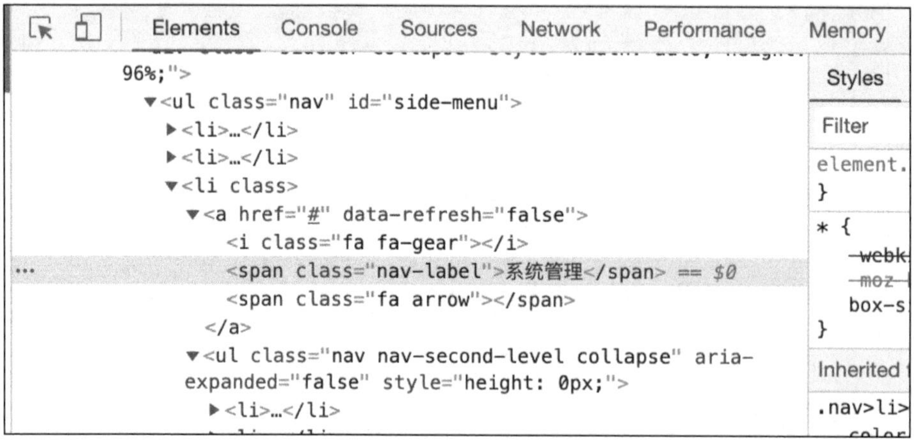

图 6.11

7. 岗位管理超链接元素

"岗位管理"是左侧导航栏二级菜单项，是一个超链接元素，其父级菜单项是"系统管理"，元素细节如图 6.12 所示。可以通过 XPath 定位方法来定位元素，值为 "//*[@id="side-menu"]/li[3]/ul/li[5]/a"。

图 6.12

岗位管理页面，如图 6.4 所示。

8. 新增超链接元素

新增超链接元素细节如图 6.13 所示，可以通过 XPath 定位方法来定位元素，值为 "//*[@id="toolbar"]/a[1]"。

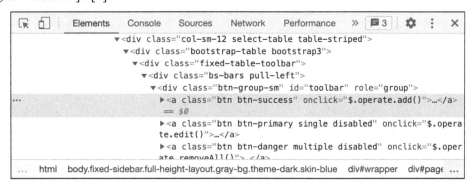

图 6.13

添加岗位页面，如图 6.5 所示。

9. 岗位名称输入框

岗位名称输入框细节如图 6.14 所示，可以通过 name 定位方法来定位元素，值为"postName"。

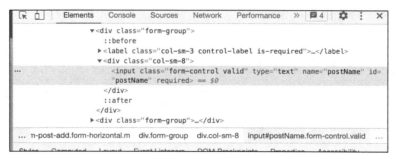

图 6.14

10. 岗位编码输入框

岗位编码输入框细节如图 6.15 所示，可以通过 name 定位方法来定位元素，值为"postCode"。

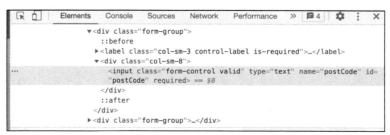

图 6.15

11. 显示顺序输入框

显示顺序输入框细节如图 6.16 所示，可以通过 name 定位方法来定位元素，值为"postSort"。

12. 备注多行文本输入区

备注多行文本输入区细节如图 6.17 所示，可以通过 name 定位方法来定位元素，值为"remark"。

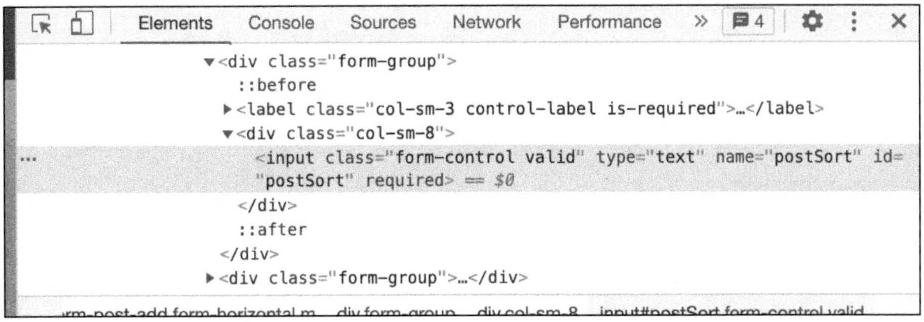

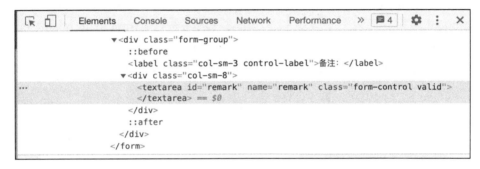

图 6.16

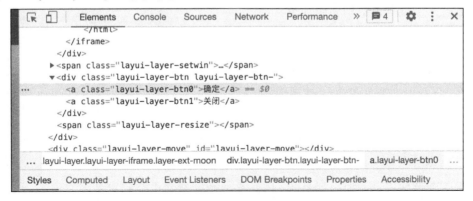

图 6.17

13. 确定超链接

确定超链接细节如图 6.18 所示,可以通过 XPath 定位方法来定位元素,值为 "//*[@id="layui-layer1"]/div[3]/a[1]"。

图 6.18

6.2.2 分层创建脚本

以上为页面元素定位的情况,在开发测试脚本之前,需要了解相关的基础知识点,便于后续进行结构化封装。

1. 在 Python 中导入模块语句

import 语句用于导入模块,原则上它可以出现在程序的任何位置。比如,import math 的作用是导入 math 模块。一次性导入多个模块的写法为 "import module1, module2,module3…";导入一个模块中的一个方法的写法为 "from math import floor";导入一个模块中的多个方法,类似于一次性导入多个模块的写法,可以用逗号分隔方法名。总体来说,import 语句导入模块时最好按照如下顺序,这样的代码逻辑比较清晰和规范。

(1) Python 标准库模块,即 Python 自带库。

(2) Python 第三方模块。

(3) 自定义模块。

2. 函数

函数是组织好的、可重复使用的、用来实现单一或相关联功能的代码段,目的是提高程序代码的重用性和可读性。Python 函数又具有一些独特的优势(可以更灵活地定义函数等),此外 Python 自身也内置了很多有用的函数,开发人员可以直接调用。

函数有六大要点,分别介绍如下。

(1) Def。

(2) 函数名。

(3) 函数体。

(4) 参数。

(5) 返回值。

(6) 两个英文符号,即小括号(),里面是参数的定义内容和冒号。

Python 函数有五大要素:def、函数名、函数体、参数、返回值。

如图 6.19 所示是一个函数的样例，这个函数实现了求和功能，其中 def sum(a,b)部分用于声明函数名"sum"和参数部分"a,b"，后面紧接着一个冒号"："，可以将其理解为声明函数的结束标志；"return a+b"是函数体，Python 语法规定函数体与函数声明部分相比要有缩进，没有缩进的话程序会报错。如果函数体结束，则缩进结束（缩进结束代表函数体结束）。示例中的函数体只有一行。

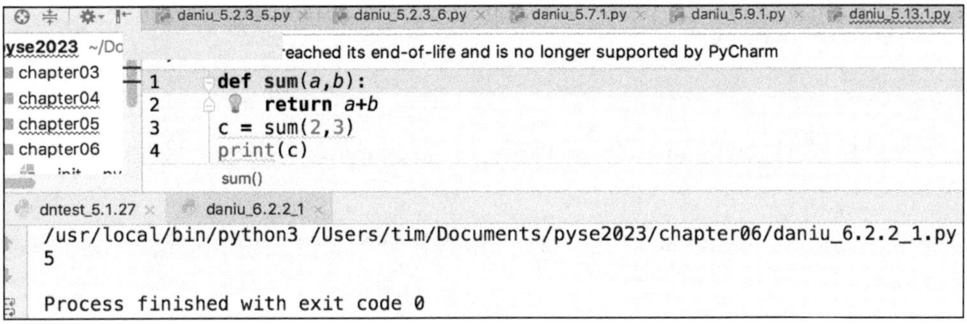

图 6.19

对于返回值来说，函数可以有返回值也可以没有返回值，如图 6.19 所示的函数是有返回值的情况，如下函数代码是无返回值的情况：

```
def sum(a,b):
    print(a+b)
print(sum(2,3))
```

Python 的函数参数模式比较复杂。以下用实例来演示各种函数参数模式。

如下函数属于位置参数类型的函数，位置参数是我们熟悉的形参，其中 x 就是一个位置参数：

```
def testFunc1(x):
    return 2 * x
```

如下函数用到了默认参数，其中 b 是默认参数，默认值为"2"。在调用函数时，如果该参数没有传入相应的值，就启用默认值。比如，调用 1：testFunc2(4)，其结果是返回"6"；调用 2：testFunc2(4,8)，其结果是返回"12"。

注意：设置参数顺序时，必须是必选参数在前，默认参数在后，否则 Python 解释器会报错；默认参数必须指向不可变对象。

```
def testFunc2(a,b = 2):
    return a + b
```

可变参数，顾名思义，在函数定义过程中，参数数量不固定，可以是 1 个、2 个，也可以是 0 个，等等。比如，在将一个 list 对象作为参数而 list 对象的值又不确定的情况下，可以使用可变参数的模式简化函数的定义。如下代码的功能是计算列表中元素的平方和，而列表元素作为参数是不确定的、可变的：

```
def testFunc3(*num_list):
    sum = 0
    for i in num_list:
        sum = sum + i * i
    return sum
```

可变参数调用 1：testFunc3([1,2,3])，用来计算列表[1,2,3]中所有元素的平方和，结果是"14"。

可变参数调用 2：testFunc3([2,3,4,5])，用来计算列表[2,3,4,5]中所有元素的平方和，结果是"54"。

关键字参数与可变参数类似，但两者的差别是，可变参数允许传入 0 个或者任意多个参数，而这些参数在调用函数时自动组成一个元组，关键字参数允许传入 0 个或任意个含参数名的参数，这些关键字参数在函数内部自动组成一个字典类型。演示实例代码如下，可通过几种函数调用来直观地了解关键字参数的特性：

```
def testFunc4(id,name,**kw):
  print('id:',id,'name:',name,'other:',kw)
```

关键字参数调用 1，执行 testFunc4('29')，控制台报错，提示缺少一个位置参数 name，结果如图 6.20 所示。

```
1  def testFunc4(id,name,**kw):
2      print('id:',id,'name:',name,'other:',kw)
3  testFunc4('29')
```

```
dntest_5.1.27    daniu_6.2.2_2
/usr/local/bin/python3 /Users/tim/Documents/pyse2023/chapter06/daniu_6.2.2_2.py
Traceback (most recent call last):
  File "/Users/tim/Documents/pyse2023/chapter06/daniu_6.2.2_2.py", line 3, in <module>
    testFunc4('29')
TypeError: testFunc4() missing 1 required positional argument: 'name'

Process finished with exit code 1
```

图 6.20

关键字参数调用 2，执行结果如图 6.21 所示，此时没有错误产生且 other 字典类型值为空。

```
def testFunc4(id,name,**kw):
    print('id:',id,'name:',name,'other:',kw)
testFunc4('29','jake')
```

```
testFunc4()
dntest_5.1.27    daniu_6.2.2_2
/usr/local/bin/python3 /Users/tim/Documents/pyse2023/chapter06/daniu_6.2.2_2.py
id: 29 name: jake other: {}

Process finished with exit code 0
```

图 6.21

关键字参数调用 3，执行结果如图 6.22 所示，other 字典类型有一个键值对{'city': 'Shanghai'}。

```
def testFunc4(id,name,**kw):
    print('id:',id,'name:',name,'other:',kw)
testFunc4('29','jake',city='shanghai')
```

```
testFunc4()
dntest_5.1.27    daniu_6.2.2_2
/usr/local/bin/python3 /Users/tim/Documents/pyse2023/chapter06/daniu_6.2.2_2.py
id: 29 name: jake other: {'city': 'shanghai'}

Process finished with exit code 0
```

图 6.22

关键字参数调用 4，执行结果如图 6.23 所示，other 字典类型有两个键值对{'city': 'Shanghai', 'age': '24'}。至此，关键字参数调用的模式和特性通过 4 个演示实例已经比较直观地展现出来，非常灵活。在使用关键字参数的时候，要注意区分不同参数模式。

```
def testFunc4(id,name,**kw):
    print('id:',id,'name:',name,'other:',kw)
testFunc4('29','jake',city='shanghai',age='24')
```

```
dntest_5.1.27    daniu_6.2.2_2
/usr/local/bin/python3 /Users/tim/Documents/pyse2023/chapter06/daniu_6.2.2_2.py
id: 29 name: jake other: {'city': 'shanghai', 'age': '24'}

Process finished with exit code 0
```

图 6.23

关键字参数允许函数调用者传入不受限制的任意关键字参数，如果要限制关键字参数的名字，可以用命名关键字参数模式进行。演示代码如下，函数 testFunc5 的关键字参数只接收“city”和“age”：

```
def testFunc5(id,name,*,city,age):
    print('id:',id,'name:',name,city,age)
```

命名关键字参数调用 1，执行结果如图 6.24 所示，关键字参数包含了“city”和“age”，而输出结果中只打印 value，没有打印 key。

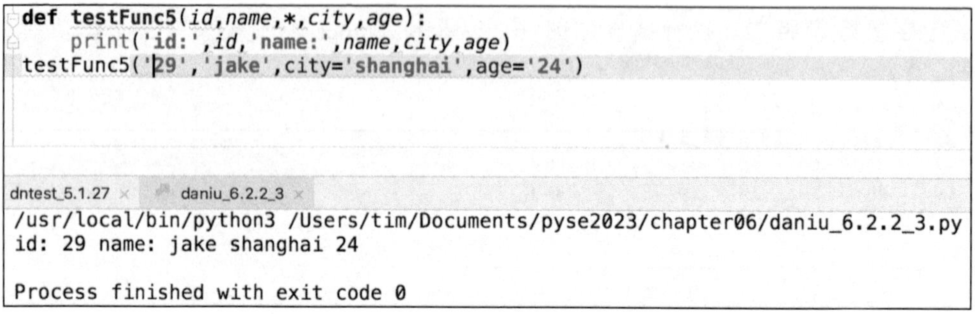
def testFunc5(id,name,*,city,age):
print('id:',id,'name:',name,city,age)
testFunc5('29','jake',city='shanghai',age='24')
dntest_5.1.27
daniu_6.2.2_3
/usr/local/bin/python3 /Users/tim/Documents/pyse2023/chapter06/daniu_6.2.2_3.py
id: 29 name: jake shanghai 24
Process finished with exit code 0

图 6.24

命名关键字参数调用 2，执行结果如图 6.25 所示，关键字参数有 3 个，分别是“city”、“age”和“job”，与函数定义有冲突（只接收“city”和“age”）。

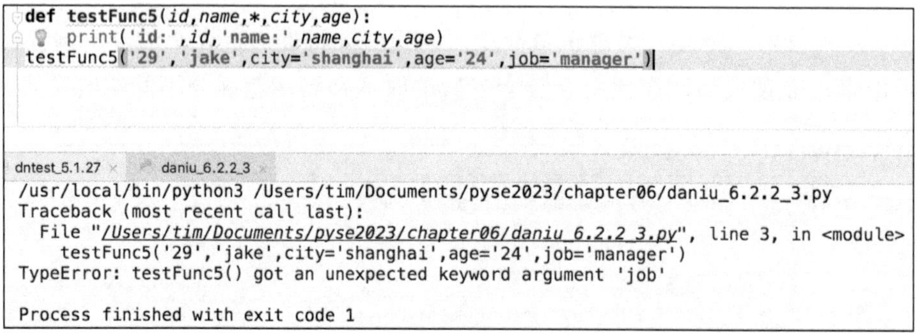
def testFunc5(id,name,*,city,age):
print('id:',id,'name:',name,city,age)
testFunc5('29','jake',city='shanghai',age='24',job='manager')
dntest_5.1.27
daniu_6.2.2_3
/usr/local/bin/python3 /Users/tim/Documents/pyse2023/chapter06/daniu_6.2.2_3.py
Traceback (most recent call last):
File "/Users/tim/Documents/pyse2023/chapter06/daniu_6.2.2_3.py", line 3, in <module>
testFunc5('29','jake',city='shanghai',age='24',job='manager')
TypeError: testFunc5() got an unexpected keyword argument 'job'
Process finished with exit code 1

图 6.25

命名关键字参数调用 3，关键字参数有 1 个，是“city”，还是与函数定义有冲突，控制台提示缺少“age”关键字参数，执行结果如图 6.26 所示。

```
def testFunc5(id,name,*,city,age):
    print('id:',id,'name:',name,city,age)
testFunc5('29','jake',city='shanghai')
```

```
/usr/local/bin/python3 /Users/tim/Documents/pyse2023/chapter06/daniu_6.2.2_3.py
Traceback (most recent call last):
  File "/Users/tim/Documents/pyse2023/chapter06/daniu_6.2.2_3.py", line 3, in <module>
    testFunc5('29','jake',city='shanghai')
TypeError: testFunc5() missing 1 required keyword-only argument: 'age'

Process finished with exit code 1
```

图 6.26

以上 Python 函数知识将用于封装登录功能，接下来开发登录功能脚本，登录功能将分 3 步完成。

第一步：完成对用户名、密码字段输入的脚本实现；第二步：完成对验证码图片的识别，以及输入正确的识别结果的脚本实现；第三步：完成"登录"按钮的实现。

第一步的实现代码如下：

```
from selenium import webdriver
from selenium.webdriver.chrome.service import Service
from selenium.webdriver.common.by import By
import time,os
driver = webdriver.Chrome()
#打开大牛测试登录页面
driver.get("http://localhost/login")
#在输入用户名之前，首先需要清空字段
driver.find_element(By.NAME,'username').clear()
driver.find_element(By.NAME,'username').send_keys("admin")
#在输入密码之前，首先需要清空字段
driver.find_element(By.NAME,'password').clear()
driver.find_element(By.NAME,'password').send_keys("admin123")
```

以上代码的执行结果如图 6.27 所示，实现了输入用户名和密码的功能。

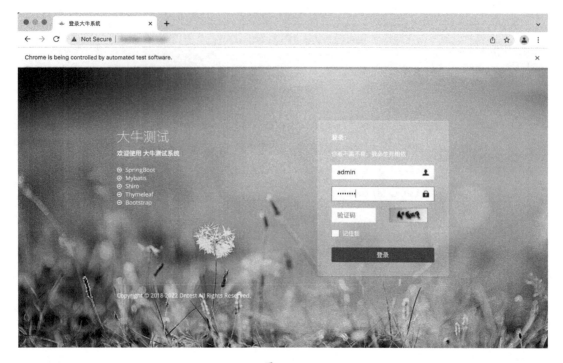

图 6.27

第二步,识别验证码,先对图片元素进行截图,将截图保存到当前目录的图片文件夹中。元素截图脚本如下:

```
#对验证码图片元素进行截图
#定义元素截图名称
filename = "capture.png"
#判定当前文件是否存在,如果存在就删除
if os.path.exists(filename):
    os.remove(filename)
elecap = driver.find_element
(By.XPATH,'//*[@id="signupForm"]/div[1]/div[2]/a/img')
#存储元素截图到 filename
elecap.screenshot(filename)
```

以上脚本执行结果如图 6.28 所示。

图 6.28

验证码图片的识别，推荐使用专业的网站，比如可以使用斐斐打码平台。访问斐斐打码平台后，先注册新用户，新用户会获得 1000 积分，如图 6.29 所示。每次图像识别都会消耗积分。

图 6.29

在斐斐打码平台官方文档中找到大牛测试系统的验证码识别类型，识别类型如图 6.30 所示，适用类型为"50100"，即简单计算题模式。

图 6.30

接下来进入开发文档选项，下载 Python 语言接口包，在下载的文件 fateadm_api.py 中，需要更改用户个人信息，如图 6.31 所示。需要在函数 TestFunc 中输入"pd_id"和"pd_key"，可以在用户中心查询这两个字段的值，如图 6.31 所示。不用填写 app_id 和 app_key，给 pred_type 赋值"50100"，给 file_name 赋值验证码的截图路径，如"capture.png"。

```
def TestFunc():
    pd_id          = "          "        #可以在用户中心查询到pd信息
    pd_key         = "                        "
    app_id         = "          "        #开发者分成用的账号,在开发者中心可以查询到
    app_key        = "                        "
    #识别类型
    #可以查看官方网站的价格页来选择具体的类型,不清楚类型的,可以咨询客服
    pred_type      = "50100"
    api            = FateadmApi(app_id, app_key, pd_id, pd_key)
    # 查询余额
    balance        = api.QueryBalcExtend()    # 直接返回余额
    # api.QueryBalc()

    # 通过文件形式识别
    file_name      = "capture.png"
    # 具体类型是多网站类型时,需要增加src_url参数,具体可参考API文档
    result     =   api.PredictFromFileExtend(pred_type,file_name)   # 直接返回识别结果
```

图 6.31

为使 TestFunc 函数直接返回验证码处理的结果,需要对 Python 软件包代码进行调整,修改 TestFunc 函数,直接返回 result 结果,调整后的代码如下:

```
def TestFunc():
    pd_id          = "135041"            #可以在用户中心查询到 pd 信息
    pd_key         = "iknN+fDXCHspi4oT0CI02Xwr"
    app_id         = "100001"            #开发者分成用的账号,在开发者中心可以查询到
    app_key        = "123456"
    #识别类型
    #可以查看官方网站的价格页来选择具体的类型,不清楚类型的,可以咨询客服
    pred_type      = "30400"
    api            = FateadmApi(app_id, app_key, pd_id, pd_key)
    #查询余额
    #balance       = api.QueryBalcExtend()    #直接返回余额
    #api.PredictExtend()
    #api.QueryBalc()
    #通过文件形式识别
    #file_name     = os.getcwd().split("common")[0]+"\\t.png"
    file_name = os.getcwd().split("common")[0]+"\\pic.png"
    result =   api.PredictFromFileExtend(pred_type,file_name)
    #直接返回识别结果
    #print(result)
    return  result
```

只需将修改后的斐斐打码源文件 fateadm_api.py 存放到 chapter06 文件夹中,获取并输入验证码功能代码如下:

```
from selenium import webdriver
from selenium.webdriver.chrome.service import Service
from selenium.webdriver.common.by import By
import time,os
import fateadm_api
#引入斐斐打码验证码Python包,源文件在/chapter06/fateadm_api.py中
#这是在macOS上执行的基于Chrome浏览器的测试,请根据实际情况修改如下的driver地址
chrome_driver_server = Service("/Users/xxx/Downloads/chromedriver")
driver  = webdriver.Chrome(service=chrome_driver_server)
driver.get("http://******.site/login")
filename = "capture.png"
if os.path.exists(filename):
    os.remove(filename)
ele1 = driver.find_element
(By.XPATH,'//*[@id="signupForm"]/div[1]/div[2]/a/img')
ele1.screenshot(filename)
#如下调用了斐斐打码的代码,直接获取识别后的验证码值
verification_code = str(fateadm_api.TestFunc())
#在页面上输入验证码
driver.find_element(By.NAME,'validateCode').send_keys(verification_code)
```

至此,大牛测试系统登录功能还剩"登录"按钮的实现,实现"登录"按钮的脚本如下:

```
driver.find_element(By.ID,'btnSubmit').click()
```

大牛测试系统登录脚本完成后,再来实现新增岗位测试脚本,这部分可以分3步进行。

(1) 在系统首页单击左侧导航栏中的"系统管理",关键代码如下:

```
driver.find_element(By.XPATH,'//*[@id="side-menu"]/li[3]/a/span[1]').click()
```

(2) 单击次级菜单"岗位管理",关键代码如下:

```
driver.find_element(By.XPATH,'//*[@id="side-menu"]/li[3]/ul/li[5]/a').click()
```

(3) 岗位管理页面新增岗位信息脚本的实现,因页面中包含iFrame,不能直接定位岗位页面信息:

```
driver.find_element(By.XPATH,'//*[@id="toolbar"]/a[1]').click()
```

因包含iFrame,需要先切换为Frame,iFrame信息如图6.32所示。

```
▼<div class="row mainContent" id="content-main">
    ::before
  ▶<iframe class="RuoYi_iframe" name="iframe0" width="100%" height="100%" data-id=
  "/system/main" src="/system/main" frameborder="0" seamless style="display: none;
  ">…</iframe>
  ▼<iframe class="RuoYi_iframe" name="iframe6" width="100%" height="100%" src="/sy
  stem/post" frameborder="0" data-id="/system/post" seamless>
    ▼#document
        <!DOCTYPE html>
      ▼<html lang="zh">
        ▶<head>…</head>
        ▼<body class="gray-bg">
          ▼<div class="container-div">
            ▼<div class="row">
                ::before
              ▶<div class="col-sm-12 search-collapse">…</div>
              ▼<div class="col-sm-12 select-table table-striped">
                ▼<div class="bootstrap-table bootstrap3">
                  ▼<div class="fixed-table-toolbar">
                    ▼<div class="bs-bars pull-left">
                      ▼<div class="btn-group-sm" id="toolbar" role="group">
                        ▼<a class="btn btn-success" onclick="$.operate.add()"> == $0
                          ▶<i class="fa fa-plus">…</i>
                            " 新增 "
                          </a>
                        ▶<a class="btn btn-primary single disabled" onclick="$.opera
                        te.edit()">…</a>
                        ▶<a class="btn btn-danger multiple disabled" onclick="$.oper
                        ate.removeAll()">…</a>
                        ▶<a class="btn btn-warning" onclick="$.table.exportExcel()">
                        …</a>
                      </div>
                    </div>
```

图 6.32

将 iFrame 切换为 Frame 的语句如下,"iframe6"是 iFrame 的 name 属性值:

```
driver.switch_to.frame("iframe6")
```

岗位名称元素在另一个 iFrame 内,iFrame name 是"layui-layer-iframe1"。但是在切换到一个新的 iFrame 之前,需要先切换回主页面,脚本代码如下:

```
driver.switch_to.default_content()
driver.switch_to.frame("layui-layer-iframe1")
driver.find_element(By.NAME,'postName').send_keys('Post1')
```

输入岗位编码:

```
driver.find_element(By.NAME,'postCode').send_keys('PostCode1')
```

输入显示顺序:

```
driver.find_element(By.NAME,'postSort').send_keys('1')
```

默认岗位状态是"正常",如果没有特殊需求,可以不处理岗位状态字段。

输入备注:

```
driver.find_element(By.NAME,'remark').send_keys("remark text 1")
```

单击"确定"按钮,在操作"确定"按钮时需要将 driver 切换回主页面,脚本如下:

```
driver.switch_to.default_content()
driver.find_element(By.XPATH,'//*[@id="layui-layer1"]/div[3]/a[1]').click()
```

6.3 项目代码总结

根据以上过程,本实战项目的代码如下,在后续的章节中会逐步深入优化代码,让读者对项目实战有一个递进学习的过程,以掌握更多项目落地细节。

大牛测试系统登录代码为 login.py,需要将涉及的验证码识别相关软件包 fateadm_api.py 存放在 chapter06 文件夹中。

```python
from selenium import webdriver
from selenium.webdriver.chrome.service import Service
from selenium.webdriver.common.by import By
import time,os
import fateadm_api
#这是在 macOS 上执行的基于 Chrome 浏览器的测试,请根据实际情况修改如下的 driver 地址
chrome_driver_server = Service("/Users/xxx/Downloads/chromedriver")
driver = webdriver.Chrome(service=chrome_driver_server)
driver.get("http://******.site/login")
filename = "capture.png"
if os.path.exists(filename):
    os.remove(filename)
ele1 = driver.find_element(By.XPATH,'//*[@id="signupForm"]/div[1]/div[2]/ a/img')
ele1.screenshot(filename)
driver.find_element(By.NAME,'username').clear()
driver.find_element(By.NAME,'username').send_keys("admin")
driver.find_element(By.NAME,'password').clear()
driver.find_element(By.NAME,'password').send_keys("admin123")
#如下调用了斐斐打码的代码,直接获取识别后的验证码值
verification_code = str(fateadm_api.TestFunc())
```

```python
#在页面上输入验证码
driver.find_element(By.NAME,'validateCode').send_keys(verification_code)
#单击"登录"按钮
driver.find_element(By.ID,'btnSubmit').click()
```

新建岗位管理文件名为 add_position.py,详细代码如下:

```python
from selenium import webdriver
from selenium.webdriver.chrome.service import Service
from selenium.webdriver.common.by import By
import time,os
#这是在macOS上执行的基于Chrome浏览器的测试,请根据实际情况修改如下的driver地址
chrome_driver_server = Service("/Users/jason118/Downloads/chromedriver")
driver = webdriver.Chrome(service=chrome_driver_server)
driver.get("http://*******.site/index")
#单击左侧导航栏->系统管理
driver.find_element(By.XPATH,'//*[@id="side-menu"]/li[3]/a/span[1]').click()
time.sleep(5)
#单击左侧导航栏->系统管理->岗位管理
driver.find_element(By.XPATH,'//*[@id="side-menu"]/li[3]/ul/li[5]/a').click()
time.sleep(2)
#单击"新增"超链接
#切换driver到iFrame,因为"新增"超链接位于iFrame内
driver.switch_to.frame("iframe6")
time.sleep(2)
driver.find_element(By.XPATH,'//*[@id="toolbar"]/a[1]').click()
time.sleep(2)
#首先返回主页面,释放iFrame
driver.switch_to.default_content()
#再次进入新的iFrame
driver.switch_to.frame("layui-layer-iframe1")
#在添加岗位页面中,输入岗位名称
driver.find_element(By.NAME,'postName').send_keys('Post1')
#在添加岗位页面中,输入岗位编码
driver.find_element(By.NAME,'postCode').send_keys('PostCode1')
#在添加岗位页面中,输入显示顺序
driver.find_element(By.NAME,'postSort').send_keys('1')
#在添加岗位页面中,输入备注
```

```python
driver.find_element(By.NAME,'remark').send_keys("remark text 1")
#首先返回到主页面,释放iFrame
time.sleep(3)
driver.switch_to.default_content()
#在添加岗位页面中,单击"确定"按钮
driver.find_element(By.XPATH,'//*[@id="layui-layer1"]/div[3]/a[1]').click()
```

第 7 章 项目重构与代码优化

本章将对代码进行初步的结构化封装，用三层结构对脚本分层，帮助读者初步建立框架思想。

7.1 项目重构

本章继续以大牛测试系统后台项目为例，在原先的代码上做进一步优化和重构，这将有利于加深大家对项目重构的认识。项目重构通常利用抽象的方法重新组织代码，进而有效地提高代码的重用性和可维护性。

7.1.1 元素定位方法优化

元素定位方法可能会被多处代码调用，此案例中就涉及多个页面，如系统登录页面、系统管理页面等。每个页面在进行元素定位时都要使用相同的定位方法，因此需要对元素定位方法进行重构再封装。

重构代码的目的主要有两点,一是减少代码量且有效提高代码重用率,二是提高代码的可读性。重构代码主要通过定义函数来实现。

首先对脚本 login.py 进行重构。第一个函数根据元素 name 属性值来返回元素定位语句。其中 name 为函数名,element 为函数参数。在函数体中返回函数定义语句,其中 name 属性值为函数传入参数 element:

```
def name(element):
    return driver.find_element(By.NAME,element)
```

第二个函数根据元素 id 属性值来返回元素定位语句:

```
def id(element):
    return driver.find_element(By.ID,element)
```

代码重构之后,测试脚本代码如下:

```
#此页面的功能是大牛测试系统的登录功能
from selenium import webdriver
from selenium.webdriver.chrome.service import Service
from selenium.webdriver.common.by import By
import time,os
import fateadm_api
chrome_driver_server = Service("./chromedriver")
driver = webdriver.Chrome(service=chrome_driver_server)
driver.get("http://localhost/login")
filename = "capture.png"
if os.path.exists(filename):
    os.remove(filename)
ele1 = driver.find_element
(By.XPATH,'//*[@id="signupForm"]/div[1]/div[2]/a/img')
ele1.screenshot(filename)
def name(element):
    return driver.find_element(By.NAME,element)
def id(element):
    return driver.find_element(By.ID,element)
name('username').clear()
name('username').send_keys("admin")
#driver.find_element(By.NAME,'username').send_keys("admin")
name('password').clear()
#driver.find_element(By.NAME,'password').clear()
name('password').send_keys("admin123")
#如下调用了斐斐打码的代码,直接获取识别后的验证码值
```

```
verification_code = str(fateadm_api.TestFunc())
#在页面上输入验证码
name('validateCode').send_keys(verification_code)
#单击"登录"按钮
id('btnSubmit').click()
```

对脚本 add_position.py 进行重构,需要使用 name 和 xpath 函数,用 XPath 元素定位方法来定位函数。其中 element 为参数,是元素值。代码如下:

```
def xpath(element):
    return driver.find_element(By.XPATH,element)
```

代码重构之后,测试脚本 add_position.py 的代码如下:

```
from selenium import webdriver
from selenium.webdriver.chrome.service import Service
from selenium.webdriver.common.by import By
import time,os
chrome_driver_server = Service("./chromedriver")
driver = webdriver.Chrome(service=chrome_driver_server)
driver.get("http://localhost/login")
#登录功能,可先用手动方式输入
time.sleep(15)
def name(locator):
    return driver.find_element(By.NAME,locator)
def xpath(locator):
    return driver.find_element(By.XPATH,locator)
#单击左侧导航栏->系统管理
xpath('//*[@id="side-menu"]/li[3]/a/span[1]').click()
time.sleep(5)
#单击左侧导航栏->系统管理->岗位管理
xpath('//*[@id="side-menu"]/li[3]/ul/li[5]/a').click()
time.sleep(2)
#单击"新增"超链接
#切换 driver 到 iFrame,因为"新增"超链接位于 iFrame 内
driver.switch_to.frame("iframe6")
time.sleep(2)
xpath('//*[@id="toolbar"]/a[1]').click()
time.sleep(2)
#首先返回主页面,释放 iFrame
driver.switch_to.default_content()
#再次进入新的 iFrame
driver.switch_to.frame("layui-layer-iframe1")
```

```
#在添加岗位页面中,输入岗位名称
name('postName').send_keys('Post1')
#在添加岗位页面中,输入岗位编码
name('postCode').send_keys('PostCode1')
#在添加岗位页面中,输入显示顺序
name('postSort').send_keys('1')
#在添加岗位页面中,输入备注
name('remark').send_keys("remark text 1")
#首先返回主页面,释放iFrame
time.sleep(3)
driver.switch_to.default_content()
#在添加岗位页面中,单击"确定"按钮
xpath('//*[@id="layui-layer1"]/div[3]/a[1]').click()
time.sleep(5)
```

7.1.2 新增岗位优化

7.1.1 节中添加的岗位信息脚本代码（add_position.py）第一次运行时可以运行成功，但从第二次开始却出现错误。错误具体如图 7.1 所示。所以从脚本代码的健壮性角度考虑，需要对相关代码进行优化。

图 7.1

从图 7.1 的错误分析，需要对岗位名称和岗位编码的输入值进行优化。优化的细节如图 7.2 所示。

```
Chapter9 > 🐍 add_position.py > ...
31    time.sleep(2)
32    xpath('//*[@id="toolbar"]/a[1]').click()
33    time.sleep(2)
34
35    #首先返回到主页，释放iFrame
36    driver.switch_to.default_content()
37    #再次进入另外的新的iFrame
38    driver.switch_to.frame("layui-layer-iframe1")
39    #在添加岗位页面，输入岗位名称
40    #对岗位名称进行优化：
41    ts = time.time()
42    postName = 'Post'+ str(ts)
43    name('postName').send_keys(postName)
44
45    #在添加岗位页面，输入岗位编码
46    #对岗位编码进行优化：
47    ts = time.time()
48    postCode = 'PostCode'+ str(ts)
49    name('postCode').send_keys(postCode)
50    #在添加岗位页面，输入显示顺序
51    name('postSort').send_keys('1')
52
53    #在添加岗位页面，输入备注
54    name('remark').send_keys("remark text 1")
55
```

图 7.2

优化后的脚本（add_position.py）如下：

```python
from selenium import webdriver
from selenium.webdriver.chrome.service import Service
from selenium.webdriver.common.by import By
import time,os
chrome_driver_server = Service("./chromedriver")
driver = webdriver.Chrome(service=chrome_driver_server)
driver.get("http://localhost/login")
time.sleep(15)
def name(locator):
    return driver.find_element(By.NAME,locator)
def xpath(locator):
    return driver.find_element(By.XPATH,locator)
#单击左侧导航栏->系统管理
xpath('//*[@id="side-menu"]/li[3]/a/span[1]').click()
time.sleep(5)
```

```
#单击左侧导航栏->系统管理->岗位管理
xpath('//*[@id="side-menu"]/li[3]/ul/li[5]/a').click()
time.sleep(2)
#单击"新增"超链接
#切换driver到iFrame,因为"新增"超链接位于iFrame内
driver.switch_to.frame("iframe6")
time.sleep(2)
xpath('//*[@id="toolbar"]/a[1]').click()
time.sleep(2)
#首先返回主页面,释放iFrame
driver.switch_to.default_content()
#再次进入新的iFrame
driver.switch_to.frame("layui-layer-iframe1")
#在添加岗位页面中,输入岗位名称
#对岗位名称进行优化
ts = time.time()
postName = 'Post'+ str(ts)
name('postName').send_keys(postName)
#在添加岗位页面中,输入岗位编码
#对岗位编码进行优化
ts = time.time()
postCode = 'PostCode'+ str(ts)
name('postCode').send_keys(postCode)
#在添加岗位页面中,输入显示顺序
name('postSort').send_keys('1')
#在添加岗位页面中,输入备注
name('remark').send_keys("remark text 1")
#首先返回主页面,释放iFrame
time.sleep(3)
driver.switch_to.default_content()
#在添加岗位页面中,单击"确定"按钮
xpath('//*[@id="layui-layer1"]/div[3]/a[1]').click()
```

7.1.3 代码分层优化

在本节中,将继续优化以上代码。通过观察发现,上面的代码函数与测试代码位于同一个文件中。随着自动化测试的深入,测试的内容和范围会逐步增加,这样的编码方式不

利于提高代码的可扩展性和可维护性。

为了更好地理解代码分层的理念，笔者将对同样的项目逐步进行深入挖掘和优化。如图 7.3 所示为初步代码分层后的代码结构图。文件"fateadm_api.py"为验证码图片处理源文件，文件"add_position.py""login.py"为测试脚本代码，文件"functions.py"主要用于存放常用的基础方法等。

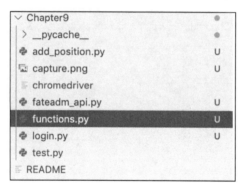

图 7.3

也可以在常用的基础方法中新增返回 driver 对象的函数 return_driver，这样做可以减少每次调用 driver 对象的代码，具体代码如下：

```
from selenium import webdriver
from selenium.webdriver.chrome.service import Service
from selenium.webdriver.common.by import By
import time
chrome_driver_server = Service("./chromedriver")
driver  = webdriver.Chrome(service=chrome_driver_server)
'''
函数 return_driver 用于返回 driver 对象
'''
def return_driver():
    return driver
'''
函数 name 使用 name 定位方法来定位元素
'''
def name(element):
    return driver.find_element(By.NAME, element)
'''
```

```
函数 id 使用 id 定位方法来定位元素
'''
def id(element):
    return driver.find_element(By.ID, element)
'''
函数 xpath 使用 XPath 定位方法来定位元素
'''
def xpath(element):
    return driver.find_element(By.XPATH,element)
```

其中测试脚本代码文件 login.py 更新如下：

```
import time,os
import fateadm_api
from functions import return_driver, id, name,xpath
driver = return_driver()
driver.get("http://localhost/login")
filename = "capture.png"
if os.path.exists(filename):
    os.remove(filename)
ele1 = xpath('//*[@id="signupForm"]/div[1]/div[2]/a/img')
ele1.screenshot(filename)
name('username').clear()
name('username').send_keys("admin")
#driver.find_element(By.NAME,'username').send_keys("admin")
name('password').clear()
#driver.find_element(By.NAME,'password').clear()
name('password').send_keys("admin123")
#如下调用了斐斐打码的代码，直接获取识别后的验证码值
verification_code = str(fateadm_api.TestFunc())
#在页面中输入验证码
name('validateCode').send_keys(verification_code)
#单击"登录"按钮
id('btnSubmit').click()
```

其中测试脚本代码文件 add_position.py 更新如下：

```
import time,os
from functions import return_driver,name,id,xpath
driver = return_driver()
```

```python
driver.get("http://localhost/login")
time.sleep(15)
#单击左侧导航栏->系统管理
xpath('//*[@id="side-menu"]/li[3]/a/span[1]').click()
time.sleep(5)
#单击左侧导航栏->系统管理->岗位管理
xpath('//*[@id="side-menu"]/li[3]/ul/li[5]/a').click()
time.sleep(2)
#单击"新增"超链接
#切换driver到iFrame,因为"新增"超链接位于iFrame内
driver.switch_to.frame("iframe6")
time.sleep(2)
xpath('//*[@id="toolbar"]/a[1]').click()
time.sleep(2)
#首先返回主页面,释放iFrame
driver.switch_to.default_content()
#再次进入新的iFrame
driver.switch_to.frame("layui-layer-iframe1")
#在添加岗位页面中,输入岗位名称
#对岗位名称进行优化
ts = time.time()
postName = 'Post'+ str(ts)
name('postName').send_keys(postName)
#在添加岗位页面中,输入岗位编码
#对岗位编码进行优化
postCode = 'PostCode'+ str(ts)
name('postCode').send_keys(postCode)
#在添加岗位页面中,输入显示顺序
name('postSort').send_keys('1')
#在添加岗位页面中,输入备注
name('remark').send_keys("remark text 1")
#首先返回主页面,释放iFrame
time.sleep(3)
driver.switch_to.default_content()
#在添加岗位页面中,单击"确定"按钮
xpath('//*[@id="layui-layer1"]/div[3]/a[1]').click()
```

7.1.4 三层架构

本节将继续对自动化代码进行重构,以上代码仍有弱点,且不够清晰明了。继续将分层优化的思想进行到底,首先需要对之前的代码结构做一些调整,便于自动化测试项目的管理和维护,也能降低项目的维护成本等。

如图 7.4 所示给出了项目重构的三层架构示意图。

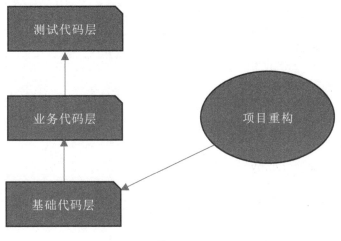

图 7.4

基于以上原则,我们可以将与 Selenium WebDriver 相关的配置、常用的基础函数等代码加入 functions.py 这个存放常用基础方法的文件中,以便其他模块调用,这属于基础代码层。

对于业务代码层,可以将之前的测试代码根据功能模块等进行分拆。比如,在此项目中,可以将添加新岗位的脚本代码抽象成一个单独的脚本文件,如 add_position.py。同理,对于系统登录业务也是适用的。它们都属于业务代码层。假如后续想添加验证岗位信息脚本,可以新增测试脚本文件 verify_position.py。

业务代码层 login.py 代码文件优化如下:

```
import time,os
import fateadm_api
from functions import return_driver, id, name,xpath
def login():
```

```python
driver = return_driver()
driver.get("http://localhost/login")
filename = "capture.png"
if os.path.exists(filename):
    os.remove(filename)
ele1 = xpath('//*[@id="signupForm"]/div[1]/div[2]/a/img')
ele1.screenshot(filename)
name('username').clear()
name('username').send_keys("admin")
name('password').clear()
name('password').send_keys("admin123")
#如下调用了斐斐打码的代码,直接获取识别后的验证码值
verification_code = str(fateadm_api.TestFunc())
#在页面中输入验证码
name('validateCode').send_keys(verification_code)
#单击"登录"按钮
id('btnSubmit').click()
```

业务代码层 add_position.py 代码文件优化如下:

```python
import time,os
from functions import return_driver,name,id,xpath
def add_position():
    driver = return_driver()
    driver.get("http://*******.site/index")
    time.sleep(15)
    #单击左侧导航栏->系统管理
    xpath('//*[@id="side-menu"]/li[3]/a/span[1]').click()
    time.sleep(5)
    #单击左侧导航栏->系统管理->岗位管理
    xpath('//*[@id="side-menu"]/li[3]/ul/li[5]/a').click()
    time.sleep(2)
    #单击"新增"超链接
    #切换driver到iFrame,因为"新增"超链接位于iFrame内
    driver.switch_to.frame("iframe6")
    time.sleep(2)
    xpath('//*[@id="toolbar"]/a[1]').click()
    time.sleep(2)
    #首先返回主页面,释放iFrame
```

```
driver.switch_to.default_content()
#再次进入新的iFrame
driver.switch_to.frame("layui-layer-iframe1")
#在添加岗位页面中,输入岗位名称
#对岗位名称进行优化
ts = time.time()
postName = 'Post'+ str(ts)
name('postName').send_keys(postName)
#在添加岗位页面中,输入岗位编码
#对岗位编码进行优化
ts = time.time()
postCode = 'PostCode'+ str(ts)
name('postCode').send_keys(postCode)
#在添加岗位页面中,输入显示顺序
name('postSort').send_keys('1')
#在添加岗位页面中,输入备注
name('remark').send_keys("remark text 1")
#首先返回主页面,释放iFrame
time.sleep(3)
driver.switch_to.default_content()
#在添加岗位页面中,单击"确定"按钮
xpath('//*[@id="layui-layer1"]/div[3]/a[1]').click()
time.sleep(5)
```

在测试代码层,可以新建测试文件 test_system.py,代码如下,实现大牛测试系统登录和新增岗位信息。可以看出测试代码非常简单:

```
from login import login
from add_position import add_position
#测试权限系统登录场景
login()
add_position()
```

代码重构是一个持续的过程,代码需要持续迭代。以上对代码实现重构之后,三层架构模型已初现:基础代码层、业务代码层和测试代码层。测试代码越来越简单,也越来越清晰。提高代码的可读性、重用性和易扩展性对自动化项目的实施非常有帮助。在自动化测试初期更需要好好规划代码结构和思路。

7.2 代码优化

代码优化一般包含很多方面。首先，可以考虑优化框架代码，如更改底层调用；然后，可以考虑厘清项目结构、优化结构组成等，便于后期维护和推广；最后，可以考虑具体的使用，比如可以考虑项目内部提出的一些代码标准、文档标准、运维管理标准等。这些都是广义上的代码优化行为。

7.2.1 无人值守自动化

在自动化测试过程中，常会发生各种异常，为了使测试代码更加健壮，需要在自动化项目中处理这些异常。如何处理异常呢？首先需要搞清楚异常产生的原因，再对这些异常进行处理。

接下来，将通过具体的例子来说明处理异常的重要性，以及处理这些异常的常用方法。

示例代码如下：

```
a = 10
b = 0
print(a/b)
```

当代码执行到第 3 行时，由于除数为 0，因此会报错，具体错误如图 7.5 所示。

```
Traceback (most recent call last):
  File "/Users/i320418/PycharmProjects/Prac1/test0307.py", line 3, in <module>
    print(a/b)
ZeroDivisionError: division by zero
```

图 7.5

如何处理和管理这些异常呢？可以用 Python 的 try 语句来捕捉异常，try except 的作用是屏蔽异常，以保证后面的代码可以执行，代码改写如下：

```
try:
    a = 10
    b = 0
    print(a/b)
except:
```

```
        print("错误,除数为零")
print('done')
```

修改代码后,代码执行结果的可读性变好了,而且代码执行完成后,会打印出便于识别的错误信息,也可以使用系统自带的除数不能为零异常,代码如下:

```
try:
    a = 1 / 0
except ZeroDivisionError as e:
    print("input wrong{}".format(e))
else:
    print("没有异常执行")
finally:
    print("不论有无异常,都执行")
```

注意,没有异常则执行 else 语句,其中 finally 语句不论有无异常都会执行。

下面以一个实际项目中的例子来说明处理异常的重要性。如果把系统登录页面用户名元素的 name 属性值"username"写错,写成"usernam",那么就会发生异常。异常截图如图 7.6 所示,提示根据属性值"usernam"不能定位元素。示例代码如下:

```
from selenium import webdriver
from selenium.webdriver.chrome.service import Service
from selenium.webdriver.common.by import By
path = Service("/Users/tim/Downloads/chromedriver")
driver = webdriver.Chrome(service=path)
driver.get("http://localhost/login")
driver.find_element(By.NAME,"usernam").clear()
driver.find_element(By.NAME,"username").send_keys("admin")
driver.find_element(By.NAME,"password").clear()
driver.find_element(By.NAME,"password").send_keys("admin123")
```

```
/usr/local/bin/python3 /Users/tim/Documents/pyse2023/chapter07/login.py
Traceback (most recent call last):
  File "/Users/tim/Documents/pyse2023/chapter07/login.py", line 8, in <module>
    driver.find_element(By.NAME,"usernam").clear()
    ^^^^^^^^^^^^^^^^^^^^^^^^^^^^^^^^^^^^^^^
  File "/Library/Frameworks/Python.framework/Versions/3.11/lib/python3.11/site-packages/selenium/webdr
    return self.execute(Command.FIND_ELEMENT, {"using": by, "value": value})["value"]
           ^^^^^^^^^^^^^^^^^^^^^^^^^^^^^^^^^^^^^^^^^^^^^^^^^^^^^^^^^^^^^^^^^^^^^^^^^^^
  File "/Library/Frameworks/Python.framework/Versions/3.11/lib/python3.11/site-packages/selenium/webdr
    self.error_handler.check_response(response)
  File "/Library/Frameworks/Python.framework/Versions/3.11/lib/python3.11/site-packages/selenium/webdr
    raise exception_class(message, screen, stacktrace)
selenium.common.exceptions.NoSuchElementException: Message: no such element: Unable to locate element:
```

图 7.6

将上面的项目代码改写如下,让测试程序捕捉到异常时打印信息"username is wrong":

```
from selenium import webdriver
from selenium.webdriver.chrome.service import Service
from selenium.webdriver.common.by import By
path = Service("/Users/tim/Downloads/chromedriver")
driver = webdriver.Chrome(service=path)
driver.get("http://localhost/login")
try:
    driver.find_element(By.NAME,"usernam").clear()
except:
    print("username is wrong")
driver.find_element(By.NAME,"username").send_keys("admin")
driver.find_element(By.NAME,"password").clear()
driver.find_element(By.NAME,"password").send_keys("admin123")
```

在搭建自动化测试框架时就需要考虑异常的处理和管理。测试人员要根据项目特点和需要来重新定义异常,便于项目内部交流和自动化测试的执行。

Selenium 中常见的异常有 9 种(如表 7.1 所示),其中比较多见的是 NoSuchElementException 异常。

表 7.1

异 常	异 常 描 述
NoSuchElementException	当选择器返回元素失败时,抛出异常
ElementNotVisibleException	当要定位的元素存在于 DOM 中,但在页面上不显示、不能交互时,抛出异常
ElementNotSelectableException	当尝试选择不可选的元素时,抛出异常
NoSuchFrameException	当要切换到目标 Frame 而 Frame 不存在时,抛出异常
NoSuchWindowException	当要切换到目标窗口,而该窗口不存在时,抛出异常
TimeoutException	当代码执行超时时,抛出异常
NoSuchAttributeException	当找不到元素的属性时,抛出异常
UnexpectedTagNameException	当支持类没有获得预期的 Web 元素时,抛出异常
NoAlertPresentException	当一个意外警告出现时,抛出异常

7.2.2 等待时间优化

在实际的项目中,代码在定位页面元素的过程中有时需要等待,但在所有定位元素的操作之前都加上等待时间,那就比较麻烦了,并且不易维护。此时可以考虑智能等待,方法很简单，即可以在代码前加上全局的智能等待时间语句，如"driver.implicitly_wait(10)"，前面提到的 add_position.py 代码可以改写如下：

```python
import time,os
from functions import return_driver,name,id,xpath
def add_position():
    driver = return_driver()
    driver.get("http://localhost/login")
#避免前面的代码频繁增加 time.sleep
driver.implicitly_wait(10)
    #单击左侧导航栏->系统管理
    xpath('//*[@id="side-menu"]/li[3]/a/span[1]').click()
    #单击左侧导航栏->系统管理->岗位管理
    xpath('//*[@id="side-menu"]/li[3]/ul/li[5]/a').click()
    #单击"新增"超链接
    #切换 driver 到 iFrame，因为"新增"超链接位于 iFrame 内
    driver.switch_to.frame("iframe6")
    xpath('//*[@id="toolbar"]/a[1]').click()
    #首先返回主页面，释放 iFrame
    driver.switch_to.default_content()
    #再次进入新的 iFrame
    driver.switch_to.frame("layui-layer-iframe1")
    #在添加岗位页面中，输入岗位名称
    #对岗位名称进行优化
    ts = time.time()
    postName = 'Post'+ str(ts)
    name('postName').send_keys(postName)
```

第 8 章 数据驱动测试

数据驱动测试是自动化测试领域主流的设计模式之一,也是高级自动化测试工程师必备的技能。数据驱动框架是一种自动化测试框架,实现同一套脚本使用不同的测试数据,测试数据和测试脚本完全分离,便于测试脚本的维护和扩展。

如测试登录操作时,需要多个用户登录,验证系统登录功能是否正常。数据驱动测试的一般步骤如下。

(1)编写脚本,脚本需要有可扩展性且支持从对象、文件或者数据库中读取测试数据。

(2)在文件或者数据库等外部介质中编写测试数据。

(3)运行脚本实现循环调用介质中的数据。

(4)验证自动化测试结果。

使用数据驱动框架,需要掌握 Python 对文件的基本操作,本章将详细讲解文件的操作方法。

8.1　一般文件操作

8.1.1　文本文件操作

Python 对文本文件的读取主要有 3 种模式。

1. read 函数模式

该模式下会一次性读取文件，缺点是文件较大时会占用更多的内存。以“dn.txt”文件来进行测试，如图 8.1 所示为内容截图。

dn.txt
dn.txt
1
daniu automation test

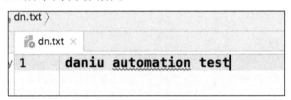

图 8.1

测试代码如下：

```
f = open("w.txt",'r')
print(f.read(2))
```

测试结果如图 8.2 所示，read 函数也支持添加参数，如输入参数 2 表示只输出前两个字符“da”，返回值为字符串类型，“r”为对文件执行读操作。

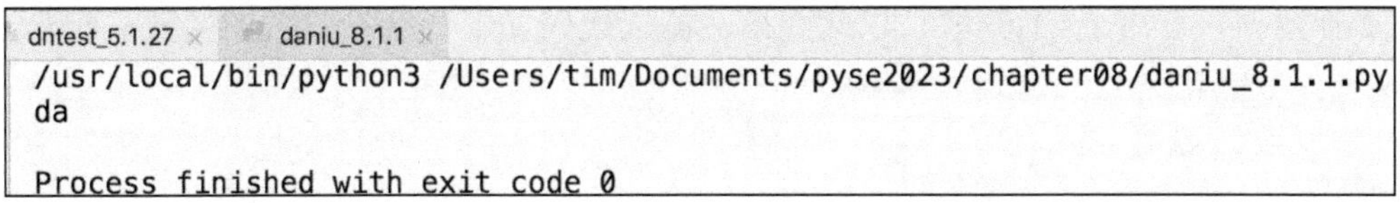

dntest_5.1.27
daniu_8.1.1
/usr/local/bin/python3 /Users/tim/Documents/pyse2023/chapter08/daniu_8.1.1.py
da
Process finished with exit code 0

图 8.2

2. readline 函数模式

该模式的特点是占用内存小、逐行读取、读取速度慢。我们以读取如图 8.3 所示的文本文件内容为例，执行结果如图 8.4 所示。测试代码如下：

```
f = open("w.txt",'r')
print(f.readline())
print(f.readline())
```

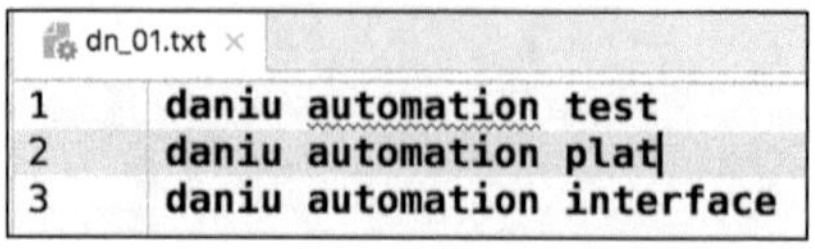

```
dn_01.txt ×
1    daniu automation test
2    daniu automation plat
3    daniu automation interface
```

图 8.3

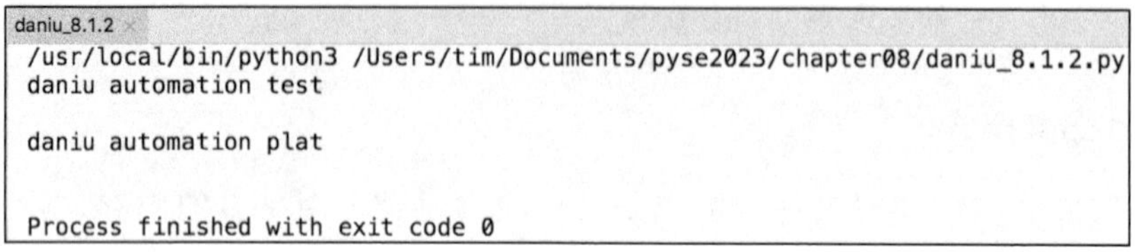

```
daniu_8.1.2 ×
/usr/local/bin/python3 /Users/tim/Documents/pyse2023/chapter08/daniu_8.1.2.py
daniu automation test

daniu automation plat

Process finished with exit code 0
```

图 8.4

3. readlines 函数模式

readlines 函数的作用是一次性读取文本内容，并将结果存储在列表中。该模式的特点是读取速度快、占用内存大。以上面的"dn_01.txt"为例，改成用 readlines 函数读取之后，执行结果如图 8.5 所示。

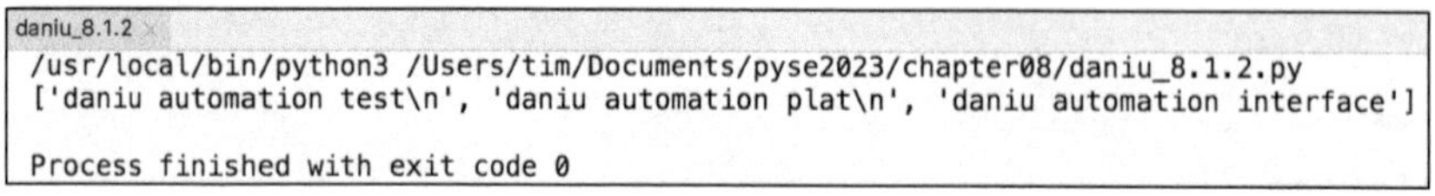

```
daniu_8.1.2 ×
/usr/local/bin/python3 /Users/tim/Documents/pyse2023/chapter08/daniu_8.1.2.py
['daniu automation test\n', 'daniu automation plat\n', 'daniu automation interface']
Process finished with exit code 0
```

图 8.5

如果要读取第 1 行数据，可用 txt[0]获取，与获取列表的第一个元素类似；如果要读取第 2 行数据，可用 txt[1]获取，以此类推。测试代码如下，执行结果如图 8.6 所示。

```
f = open("dn_01.txt")
txt = f.readlines()
#获取 readlines 返回对象的类型
print(type(txt))
#获取返回对象的值
print(txt)
#获取列表对象的第 1 个元素
print(txt[0])
```

图 8.6

文本文件写操作，以空文本文件"test.txt"为例进行测试，"w"为写操作，脚本代码如下：

```
f=open('test.txt','w')
f.write('test')
f.writelines('test daniu')
f.write('plat')
```

代码执行后，在 test.txt 文件中添加了如图 8.7 所示的字符串"testtest daniuplat"。

图 8.7

8.1.2　CSV 文件操作

CSV 的英文全称是 Comma Separated Values。实际项目中会将部分文件或测试数据存储在 CSV 文件中，因此我们要熟练掌握处理 CSV 文件的技能。

在处理 CSV 文件时，首先需要导入 CSV 模块，语句为"import csv"。例如，业务场景是读取 test.csv 文件，并打印第 1 列数据，如图 8.8 所示。

图 8.8

测试代码如下,执行结果如图 8.9 所示,控制台成功输出 CSV 文件中的第 1 列内容:

```
import csv
csvfile = "test.csv"
c = csv.reader(open(csvfile,'r'))
for cs in c:
    print(cs[0])
```

```
csv_read
/usr/local/bin/python3 /Users/tim/Documents/pyse2023/chapter08/csv_read.py
selenium
appium
interface
php
```

图 8.9

在实际测试过程中,经常要获取 CSV 文件中的各个单元格的数据。以下演示如何输出 CSV 文件中各个单元格的数据,代码如下:

```
import csv
csvfile = "test.csv"
c = csv.reader(open(csvfile,"r"))
#获取 c 对象的类型
print(type(c))
for cs in c:
    for i in range(len(cs)):
        #print(type(cs[i]))
        print(cs[i])
```

执行结果如图 8.10 所示。

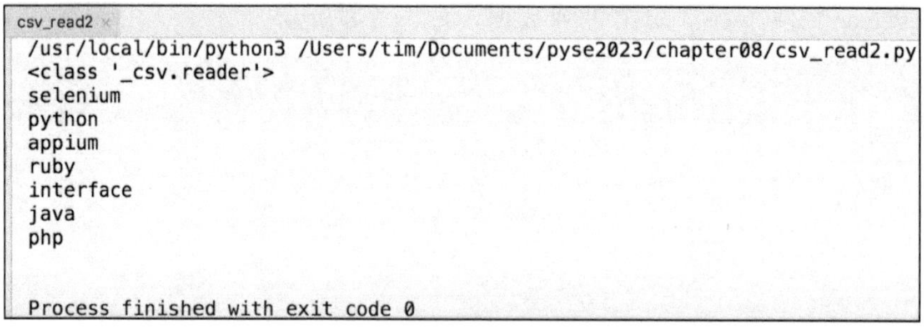

```
csv_read2
/usr/local/bin/python3 /Users/tim/Documents/pyse2023/chapter08/csv_read2.py
<class '_csv.reader'>
selenium
python
appium
ruby
interface
java
php

Process finished with exit code 0
```

图 8.10

8.1.3 Excel 文件操作

8.1.2 节介绍了处理 CSV 文件的方法，本节开始介绍处理 Excel 文件的方法。CSV 和 Excel 文件都可以用微软 Excel 打开，两者有哪些区别呢？

- Excel 文件是二进制文件，以工作簿的形式管理工作表；而 CSV 文件是文本文件，文本以逗号分隔。

- Excel 文件的功能更强大，不仅能够存储数据，而且能够嵌入处理数据的公式。而 CSV 文件相对简单很多，它只是一个普通的文本文件，并不包含格式、公式和宏命令等。

- Excel 文件不能被文本编辑器打开，而 CSV 文件可以被文本编辑器打开。

- 从编程语言的角度进行分析，当处理、解析这两种文件时，Excel 文件比 CSV 文件要复杂，且会花费更多的时间。

1. 读取 Excel 文件

Python 要读取 Excel 文件，需要先安装 xlrd 库。在命令行窗口中运行"pip install xlrd"命令，如图 8.11 所示。也可通过离线包安装，离线包直接在 Python 官网下载，我们下载当前最新的 2.0.1 版本（截至本书完稿时），如图 8.12 所示。

安装完 xlrd 库之后，便可对 Excel 文件进行处理。以读取 Excel 文件 test.xlsx 为例，如图 8.13 所示，打开 Excel 文件可以直接用库中提供的 open_workbook 方法。

注意：新版 xlrd 库因安全问题不支持 xlsx 格式，可以降低版本安装，如安装 1.2.0 版本（本项目采用 1.2.0 版本），或选择安装其他库，如 openpyxld，都支持 xlsx 格式。

```
timdeMacBook-Pro:~ tim$ pip install xlrd
Collecting xlrd
  Using cached xlrd-2.0.1-py2.py3-none-any.whl (96 kB)
Installing collected packages: xlrd
Successfully installed xlrd-2.0.1
```

图 8.11

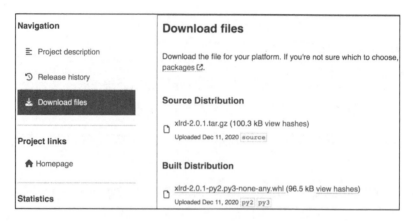

图 8.12

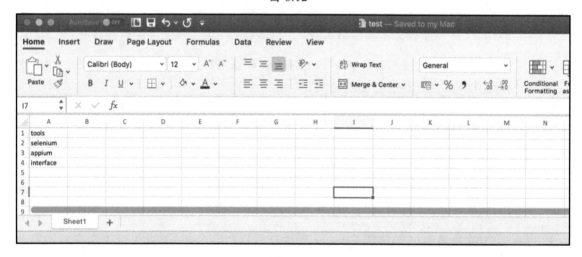

图 8.13

获取 sheet 名称的方法有 3 种，介绍如下。

（1）通过 sheets 方法获取，获取名称为"Sheet1"表的方式是 sheets()[0]。

（2）通过 sheet 名称获取，获取名称为"Sheet1"表的方式是 sheet_by_name('Sheet1')。

（3）通过 sheet 索引获取，获取名称为"Sheet1"表的方式是 sheet_by_index(0)。

下面介绍几种常用的读取 Excel 表格的方法：nrows 方法用于获取总行数，ncols 方法用于获取总列数，row_values 方法用于获取单元格数据。具体的实现方法如下，执行结果如图 8.14 所示：

```python
import xlrd
xls = xlrd.open_workbook('test.xlsx')
#sheet = xls.sheets()[0]  #通过 sheets 方法获取表格
#sheet = xls.sheet_by_name('Sheet1')  #通过 sheet_by_name 方法获取表格
sheet = xls.sheet_by_index(0)  #通过 sheet_by_index 方法获取表格
print(sheet.nrows)  #打印表格总行数
print(sheet.ncols)  #打印表格总列数
print(sheet.row_values(1)[0])  #打印表格第 2 行第 1 列的单元格数据
```

```
excel_read
/usr/local/bin/python3 /Users/tim/Documents/pyse2023/chapter08/excel_read.py
4
1
selenium

Process finished with exit code 0
```

图 8.14

更进一步地，若想获取所有的数据，可以用循环的方法，将行列数设置为变量，嵌套循环读取 Excel 表格中的单元格数据，执行结果如图 8.15 所示。代码如下：

```python
import xlrd
xls = xlrd.open_workbook('test.xlsx')
sheet = xls.sheet_by_index(0)
print(sheet.nrows)
print(sheet.ncols)
for r in range(sheet.nrows):
    for c in range(sheet.ncols):
        print(sheet.row_values(r)[c])
```

```
excel_read_all
/usr/local/bin/python3 /Users/tim/Documents/pyse2023/chapter08/excel_read_all.py
4
1
tools
selenium
appium
interface

Process finished with exit code 0
```

图 8.15

2. 写入 Excel 文件

执行 Excel 文件的写操作，需要安装 Python 的 xlwt 库，它的安装方式与安装 xlrd 库的类似，在命令行窗口中执行命令"pip install xlwt"，安装过程如图 8.16 所示。也可以通过离线包的形式来安装。

```
timdeMacBook-Pro:~ tim$ pip install xlwt
Collecting xlwt
  Using cached xlwt-1.3.0-py2.py3-none-any.whl (99 kB)
Installing collected packages: xlwt
Successfully installed xlwt-1.3.0
```

图 8.16

Excel 写操作有两种常用的方法，add_sheet 方法用于增加工作表，write 方法用于向单元格中写入数据，该方法有 3 个参数（行、列、具体值），示例代码如下，执行结果如图 8.17 所示。

```python
import xlwt
wb = xlwt.Workbook()
sheet = wb.add_sheet(u'测试')
sheet.write(0,0,"automation")
sheet.write(0,1,"selenium course")
wb.save('automate1.xls')
```

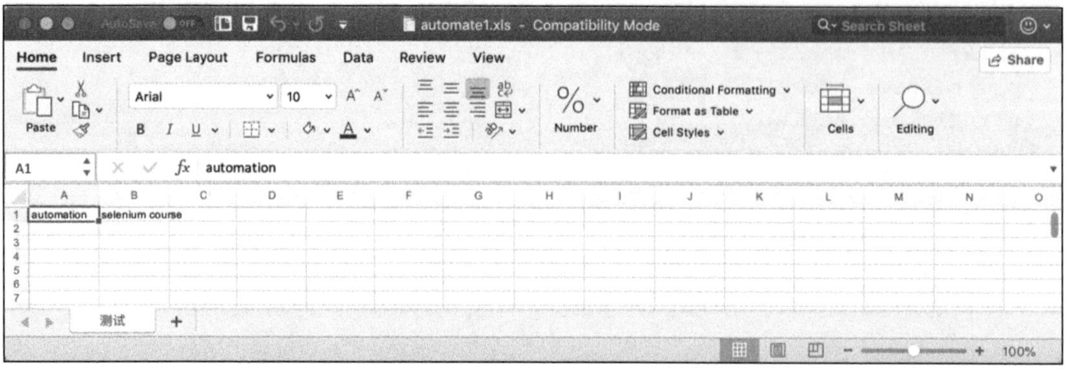

图 8.17

8.1.4 JSON 文件操作

JSON 是一种轻量级的数据交换格式，它通过一种完全独立于编程语言的文本格式来存储和展示数据。JSON 的特点是，不仅可读性强，而且也有利于机器解析和生成，其一般用于提升网络传输速率，常用于接口开发。

Python 处理 JSON 格式的数据前,需要导入 JSON 类库,命令如"import json"。

JSON 类库有两个比较重要的方法。

(1) dumps 方法:将 Python 对象编码成 JSON 字符串。

(2) loads 方法:将 JSON 字符串编码成 Python 对象。

dumps 方法示例代码如下:

```
import json
json_data = {'j1' : 1, 'j2' : 2, 'j3' : 3, 'j4' : 4}
json_1 = json.dumps(json_data)
print(json_1)
print(type(json_1))
#打印 Python 的字典元素
dict_data = {'j1' : 1, 'j2' : 2, 'j3' : 3, 'j4' : 4}
print(dict_data)
```

代码执行结果如图 8.18 所示,JSON 格式数据字符串使用双引号包裹,而字典元素字符串使用单引号包裹。

```
json_dumps ×
/usr/local/bin/python3 /Users/tim/Documents/pyse2023/chapter08/json_dumps.py
{"j1": 1, "j2": 2, "j3": 3, "j4": 4}
<class 'str'>
{'j1': 1, 'j2': 2, 'j3': 3, 'j4': 4}

Process finished with exit code 0
```

图 8.18

loads 方法示例代码如下:

```
#解码 JSON 数据,将其转化为字典类型数据
import json
json_data1 = '{"j1": 1, "j2": 2, "j3": 3, "j4": 4}'
text_json = json.loads(json_data1)
print(text_json)
print(type(text_json))
```

如图 8.19 所示,上例返回的数据是字典类型的数据,通过转换表倒推发现,对应 JSON 中的数据类型应该是 object 类型。之前通过 dumps 方法得到的是"<class 'str'>"类型,它也是 object 类型之一。

```
json_loads
/usr/local/bin/python3 /Users/tim/Documents/pyse2023/chapter08/json_loads.py
{'j1': 1, 'j2': 2, 'j3': 3, 'j4': 4}
<class 'dict'>

Process finished with exit code 0
```

图 8.19

JSON 数据类型与 Python 数据类型转换如表 8.1 所示。

表 8.1

JSON 数据类型	Python 数据类型	JSON 数据类型	Python 数据类型
object	dict	number (real)	float
array	list	TRUE	TRUE
string	unicode	FALSE	FALSE
number (int)	int, long	null	None

接下来,举例说明如何读取 JSON 文件。先准备一个名为"test.json"的 JSON 文件,内容如下:

```
{"android":"appium","web":"selenium","interface":"python interface automation"}
```

代码如下:

```
import json
f = open('test.json','r')
print(json.load(f))
```

执行结果如图 8.20 所示。

```
json_load
/usr/local/bin/python3 /Users/tim/Documents/pyse2023/chapter08/json_load.py
{'android': 'appium', 'web': 'selenium', 'interface': 'python interface automation'}

Process finished with exit code 0
```

图 8.20

以上为 JSON 文件的读取操作,对 JSON 文件执行写操作的代码如下:

```
import json
f = open("daniu.json","w")
js = {'android': 'appium', 'web1': 'selenium', 'interface': 'python interface automation'}
json.dump(js,f)
```

执行成功后，会生成名为"daniu.json"的 JSON 文件，其内容如图 8.21 所示。

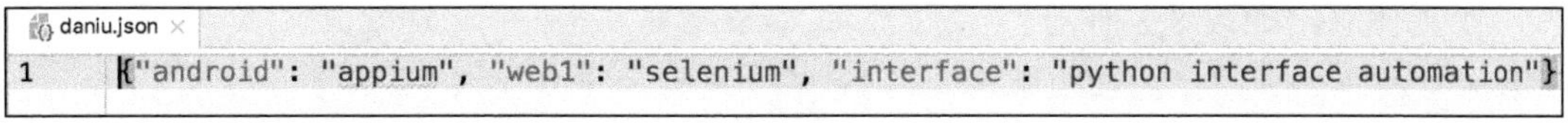

```
{"android": "appium", "web1": "selenium", "interface": "python interface automation"}
```

图 8.21

8.1.5　XML 文件操作

XML（可扩展标记语言）是互联网数据传输的重要载体，它不受系统和编程语言的限制。它是一个数据携带者且具有高级别通行证。XML 传递的具有结构化特征的数据是系统间、组件间得以沟通、交互的重要媒介之一。

编程实践中，XML 不仅可以标记数据，还可以定义数据类型等。XML 提供了统一的方法来描述和交换结构化数据。如下代码是一个名为"user.xml"的 XML 文件示例：

```xml
<?xml version="1.0" encoding="UTF-8" ?>
<users>
    <user id="1000001">
        <username>Admin1</username>
        <password>Admin1</password>
    </user>
    <user id="1000002">
        <username>Admin2</username>
        <password>Admin2</password>
    </user>
</users>
```

根据以上 XML 代码，分析 XML 文件的结果如下。

- XML 声明部分一般位于 XML 文件的第 1 行，而且声明一般包括版本号和文件字符编码格式。如上例所示，XML 文件遵循的是 1.0 版本的标准，字符编码格式为"UTF-8"。
- XML 文件的根元素必须是唯一的。它的开始标签位于文件最前面而结束标签位于文件最后。如上例中，\<users>和\</users>是文件的根元素。
- 所有的 XML 元素都必须有结束标签。
- XML 标签对大小写敏感。

- 在 XML 文件中，一些字符拥有特殊意义，不能直接使用，容易造成文件格式错误，具体总结如表 8.2 所示。

表 8.2

显示结果	描述	实体名称	显示结果	描述	实体名称
	空格		&	和号	&amp;
<	小于号	&lt;	"	引号	&quot;
>	大于号	&gt;	'	撇号	&apos;

读取 user.xml 中的用户信息，可以先用 DOM 解析 XML，再用 getElementsByTagName 方法获取 user 标签的内容。user.xml 中有两个 user，使用 list[0]可以获取第一个 user 的内容，使用 getAttribute 方法可以获取（根元素）属性，使用 getElementsByTagName 方法可以获取子标签。读取 XML 文件的代码如下：

```
import xml.dom.minidom
dom = xml.dom.minidom.parse('user.xml')
root = dom.documentElement
list = root.getElementsByTagName("user")
print(list[0].getAttribute("id"))
print(list[0].getElementsByTagName("password")[0].childNodes[0].nodeValue)
```

代码执行结果如图 8.22 所示，与预期的结果一致。

```
xml_read
/usr/local/bin/python3 /Users/tim/Documents/pyse2023/chapter08/xml_read.py
1000001
Admin1

Process finished with exit code 0
```

图 8.22

遍历 XML 文件中的所有值，代码如下：

```
import xml.dom.minidom
dom = xml.dom.minidom.parse('user.xml')
root = dom.documentElement
list = root.getElementsByTagName("user")
for l in list:
    print(l.getAttribute("id"))
    print(l.getElementsByTagName("password")[0].childNodes[0].nodeValue)
    print(l.getElementsByTagName("username")[0].childNodes[0].nodeValue)
```

代码执行结果如图 8.23 所示。

```
xml_read_all
/usr/local/bin/python3 /Users/tim/Documents/pyse2023/chapter08/xml_read_all.py
1000001
Admin1
Admin1
1000002
Admin2
Admin2

Process finished with exit code 0
```

图 8.23

以下对读取 XML 文件时使用的一些重要函数进行补充说明。

- xml.dom.minidom.parse：返回文件节点对象。

- getElementsByTagName：返回带有指定名称的所有元素的节点列表（NodeList）。

- getAttributes：返回某一元素的属性值。

8.1.6　YAML 文件操作

YAML 是一种很直观的能够被计算机识别的数据序列化格式，不仅易读且能够与脚本语言进行交互。从语法结构来看，YAML 类似于 XML，但是语法比 XML 简单。YAML（Yet Another Markup Language）翻译成中文就是"另一种标记语言"。

YAML 的用途比较广泛，在配置文件方面的应用较多。YAML 的语法结构比较简单且强大，远比 JSON 格式方便。

YAML 的基本语法如下。

- 大小写敏感。

- 行缩进时不允许使用 Tab 键，只允许使用空格键。

- 缩进的空格数没有限制，只要相同层级的元素左侧对齐就可以。

- YAML 使用缩进来表示层级关系，相同的层级元素左侧是对齐的。

- 符号"#"后的内容是注释，这点与 Python 语言一致。

Python 操作 YAML 文件时，需要先安装 PyYAML 模块，安装命令是"pip install pyyaml"，如图 8.24 所示表明 PyYAML 模块已经安装成功。

```
timdeMacBook-Pro:~ tim$ pip install pyyaml
Collecting pyyaml
  Using cached PyYAML-6.0-cp311-cp311-macosx_10_9_x86_64.whl (188 kB)
Installing collected packages: pyyaml
Successfully installed pyyaml-6.0
```

图 8.24

下面举例演示 YAML 文件的用法。

创建一个简单的 YAML 文件,名称为"config.yml",内容如下:

```
name: Jack
age: 23
children:
    name: Jason
    age: 2
    name_1: Jeff
    age_1: 4
```

读取 YMAL 文件的 Python 代码如下:

```
import yaml
file_1 = open('config.yml')
#返回一个字典对象
yml = yaml.load(file_1,Loader=yaml.FullLoader)
print(yml)
print(type(yml))
```

代码执行结果如图 8.25 所示,从代码"print(type(yml))"的执行结果可以发现,处理 YAML 文件后得到的是一个字典对象。

```
yaml_test
/usr/local/bin/python3 /Users/tim/Documents/pyse2023/chapter08/yaml_test.py
{'name': 'Jack', 'age': 23, 'children': {'name': 'Jason', 'age': 2, 'name_1': 'Jeff', 'age_1': 4}}
<class 'dict'>

Process finished with exit code 0
```

图 8.25

使用方法 yaml.dump 将一个 Python 对象转化为 YAML 文件,示例代码如下:

```
import yaml
object_1 = {'name': 'Jack', 'age': 23, 'children': {'name': 'Jason', 'age': 2, 'name_1': 'Jeff', 'age_1': 4}}
print(yaml.dump(object_1,))
```

执行结果如图 8.26 所示。

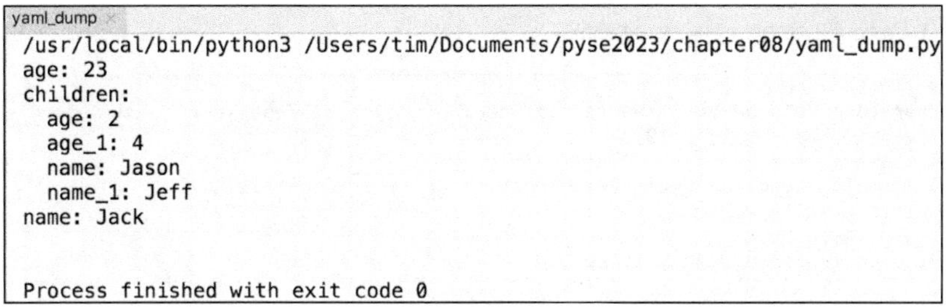

```
yaml_dump
/usr/local/bin/python3 /Users/tim/Documents/pyse2023/chapter08/yaml_dump.py
age: 23
children:
  age: 2
  age_1: 4
  name: Jason
  name_1: Jeff
name: Jack


Process finished with exit code 0
```

图 8.26

以上案例为将 YAML 格式转换为字典格式，也可以将 YAML 格式数据转化为其他 Python 对象，比如列表或者复合结构的数据（如字典和列表的复合类型数据）。示例代码如下：

```
import yaml
#将YAML格式的数据转化为Python列表类型的数据
file_2 = open('config2.yml')
yml = yaml.load(file_2,Loader=yaml.FullLoader)
print(yml)
print(type(yml))
#将YAML格式的数据转化为复合类型的数据,其中包含字典和列表数据类型
file_3 = open('config3.yml')
yml_3 = yaml.load(file_3,Loader=yaml.FullLoader)
print(yml_3)
print(type(yml_3))
```

以上代码分为两部分，第一部分是将 YAML 格式的文件转换为列表数据类型的文件。其中 YAML 数据文件内容如下，注意符号"-"与数值之间要有空格：

```
- James
- 20
- Lily
- 19
```

第二部分是将 YAML 格式的文件转换为复合类型的文件，包含列表和字典。数据文件内容如下：

```
- name: James
  age: 20
- name: Lily
  age: 19
```

最终的代码执行结果如图 8.27 所示。

```
yaml_object
/usr/local/bin/python3 /Users/tim/Documents/pyse2023/chapter08/yaml_object.py
['James', 20, 'Lily', 19]
<class 'list'>
[{'name': 'James', 'age': 20}, {'name': 'Lily', 'age': 19}]
<class 'list'>

Process finished with exit code 0
```

图 8.27

8.1.7 文件夹操作

通过前面的介绍，相信大家已经掌握了对各种文件的操作。在自动化测试过程中，不可避免地要对文件夹进行操作，接下来，将通过代码来演示文件夹相关操作，如输出文件或文件夹路径、创建或删除文件夹等：

```python
#coding=utf-8
#操作文件夹时需要导入 os 模块
import os
#打印当前执行脚本所在的目录
print(os.getcwd())
#如果当前路径下存在图片 tt.png，则返回"True"，不存在则返回"False"
print(os.path.exists('/PycharmProjects/Sstone/tt.png'))
#判断当前路径下是否存在图片 tt.png，如果存在，则返回"True"
print(os.path.isfile('/PycharmProjects/Sstone/tt.png'))
#删除多级目录
os.removedirs('/PycharmProjects/Sstone/')
#在当前目录下创建单个文件夹 test1221
os.mkdir("test1221")
#创建多级目录
os.makedirs('/PycharmProjects/Sstone/1/2/3')
```

8.2 通过 Excel 参数，实现参数与脚本的分离

在之前的项目案例中，直接在代码中维护测试数据，这种在程序中直接给代码赋值的方式俗称"hardcode"。这种方式不利于数据的修改和维护，会使程序的质量变差。

8.2.1 创建 Excel 文件，维护测试数据

我们尝试将测试数据存储到 Excel 中，创建 Excel 文件 "testdata.xlsx"，并设计 3 组登录用户名与密码数据，具体数据如图 8.28 所示。

A	B	C	D
username	password		
admin	admin123		
daniu	daniu		
daniutest	daniutest		

图 8.28

下一步用 Python 实现读取 Excel 文件的函数功能，以备测试之用，代码如下：

```
def read_excel(filename, index, cloumn):
    #运用 xlrd 库的 open 方法来打开 Excel 文件
    xls = xlrd.open_workbook(filename)
    #指定要选择的表格
    sheet = xls.sheet_by_index(index)
    #打印选定表格的行数
    print(sheet.nrows)
    #打印选定表格的列数
    print(sheet.ncols)
    #声明一个空列表 data
    data = []
    #表明用 for 循环遍历 Excel 中的第 1 列数据，然后将获得的数据加入列表 data 中
        for i in range(sheet.nrows):
            data.append(sheet.row_values(i)[0])
            print(sheet.row_values(i)[0])
            #返回列表 data
        return data
```

上述代码创建了名为 "read_excel" 的函数，并设置了 3 个参数。其中，filename 是 Excel 文件的名称，可以指定为相对路径；index 是表格的编号，如 Excel 中 Sheet1 表格的 index 值为 0；column 是表格的列号，如 A 列对应的 column 值为 0。

以上用列表的方式存储了从 Excel 读取的数据，通过观察可以看到 A 列有 3 行数据，B 列有 4 行数据。上面的 Excel 读取方法通用性较差，可通过字典进行优化，思路是把每

一列数据存储到一个列表中作为字典的 value。新的读取 Excel 文件的函数代码如下：

```python
def read_excel(filename, index):
    xls = xlrd.open_workbook(filename)
    sheet = xls.sheet_by_index(index)
    print(sheet.nrows)
    print(sheet.ncols)
    dic = {}
    for j in range(sheet.ncols):
        data = []
        for i in range(sheet.nrows):
            data.append(sheet.row_values(i)[j])
        dic[j] = data
    return dic
```

以下代码用于输出所有 Excel 文件中的第一个表格中的所有数据，代码执行结果如图 8.29 所示，与 Excel 文件中的数据一致：

```python
print(read_excel("testdata.xlsx",0))
```

```
daniu_8.2.1
/usr/local/bin/python3 /Users/tim/Documents/pyse2023/chapter08/daniu_8.2.1.py
4
2
{0: ['username', 'admin', 'daniu', 'daniutest'], 1: ['password', 'admin123', 'daniu', 'daniutest']}

Process finished with exit code 0
```

图 8.29

8.2.2　Framework Log 设置

关于日志，软件开发人员或测试人员应该都不陌生，它是追踪应用运行时所发生事件的一种方法。事件（Event）是有轻重缓急的，可以用严重等级来区分，相应的日志也用日志等级来区分。

日志非常重要，通过日志可以了解应用的运行情况，便于分析出用户偏好、习惯、操作行为等。日志的作用有以下两点。

（1）调试程序。

（2）了解软件健康状况，查看软件运行是否正常等。现在基于日志的分析统计软件有很多，如 Splunk 就是其中的佼佼者，它提供了很多日志分析、查询、统计功能，以及强大的报表定制化功能等。

不同系统或者软件有不同的日志等级定义，总结一下，常用的日志等级如下。

- DEBUG
- INFO
- WARNING
- ERROR
- ALERT
- NOTICE
- CRITICAL

日志的一般组成结构如下。

- 事件发生的时间，有些国际化软件的日志中还包含时区信息，比如 GMT 等。
- 事件发生的位置，比如事件发生时，程序执行的代码信息等。
- 事件的严重程度，也就是日志等级。
- 事件的内容，一般由开发者控制，要输出哪些内容，以及以什么样的格式输出。

一般的开发语言都会包含与日志相关的模块（功能），比如 Log4j、log4php 等，它们的功能强大，使用简单。Python 也提供了日志的标准库模块 logging。

logging 模块的日志等级设定如下。

- DEBUG，通常日志信息很详细，该等级的设定场景是问题定位和调试。
- INFO，日志信息详细程度仅次于 DEBUG，通常只记录关键信息点，用于确认软件是否按照正常的预期运行。
- WARNING，当出现某些异常信息时，系统记录的日志信息，此时不影响软件正常运行。
- ERROR，记录因更严重的问题而导致的软件运行不正常的相关日志信息，如内容溢出异常等。

- CRITICAL，当严重的错误发生而直接导致宕机、软件服务等无法使用时记录的日志信息。

日志等级从低到高依次为 DEBUG < INFO < WARNING < ERROR < CRITICAL，但是相应的日志信息量是逐步减少的。

logging 模块定义日志等级的常用方法如下。

- logging.debug(msg,*args,**kwargs)
- logging.info(msg,*args,**kwargs)
- logging.warning (msg,*args,**kwargs)
- logging.error(msg,*args,**kwargs)
- logging.critical(msg,*args,**kwargs)

以上方法的作用是创建如 DEBUG、INFO、WARNING、ERROR、CRITICAL 等日志等级的日志。此外，还有两个常用方法，作用如下。

- logging.log(level,*args,**kwargs)：用于创建特定日志等级（由 level 指定）的日志信息。
- logging.basicConfig(**kwargs)：对 root logger 进行配置，主要用于指定"日志等级""日志格式""日志输出位置/文件""日志文件的打开模式"等信息。

logging 模块的四大组件如下。

（1）loggers：提供可供程序直接调用的接口。

（2）handlers：用于将日志信息发送到指定的位置。

（3）filters：提供日志过滤功能。

（4）formatters：提供日志输出格式设定功能。

以下为简单的 logging 模块使用示例代码：

```
import logging
logging.debug("I am a debug level log.")
logging.info("I am a info level log.")
logging.warning("I am a warning level log.")
logging.error("I am a error level log.")
```

```
logging.critical(" I am a critical level log.")
```

以上示例也可以使用另一种写法，代码如下：

```
import logging
logging.log(logging.DEBUG,"I am a debug level log.")
logging.log(logging.INFO,"I am a info level log.")
logging.log(logging.WARNING,"I am a warning level log.")
logging.log(logging.ERROR,"I am a error level log.")
logging.log(logging.CRITICAL,"I am a critical level log.")
```

在控制台上打印如下结果：

```
WARNING:root:I am a warning level log.
ERROR:root:I am a error level log.
CRITICAL:root: I am a critical level log.
```

可以发现，DEBUG 和 INFO 等级的日志没有输出，这是因为 logging 模块提供的日志记录方法所使用的日志器设置的级别为 WARNING，因此只有 WARNING 等级及大于该等级的（如 ERROR、CRITICAL）日志才会输出，而等级比 WARNING 低的日志会被丢弃。

打印出来的日志信息如"WARNING:root:I am a warning level log."，各个字段的含义分别是日志等级、日志器名称和日志内容。之所以用这样的格式输出日志信息，是因为日志器中设置的是默认格式 BASIC_FORMAT，格式为"%(levelname)s:%(name)s:%(message)s"。

另外，为什么日志被打印到控制台而不是别的地方？原因是日志器中使用的是默认输出位置"sys.stderr"。

如果要改变日志输出位置，需要手动调用方法 basicConfig 进行设置。basicConfig 方法的定义为"logging.basicConfig(**kwargs)"。

该方法的参数描述如下。

- Filename：指定输出的目标文件名，用于保存日志信息。设置该参数后，便不会将日志信息打印到控制台。
- FileMode：指定日志文件的打开模式，默认为"a"，且仅在指定了 Filename 时该参数生效。
- Format：指定输出日志的格式和内容，Format 可用于输出很多有用的信息。
- DateFmt：指定日期/时间格式。
- Level：指定日志器的日志等级。

- Stream：指定日志输出目标 Stream，比如"sys.stdout""sys.stderr"。需要注意的是，不能同时提供 Stream 参数和 Filename 参数，若同时指定，可能会造成冲突和产生 ValueError 异常。
- Style：这是在 Python 3 版本之后新增的参数，用于指定 Format 格式字符串的风格，取值为"%"、"{"和"$"，默认值为"%"。
- Handlers：这是在 Python 3.6 版本之后新增的参数。如果指定该参数，它是一个创建了多个 Handler 的可迭代对象，这些 Handler 将会被添加到 root logger 中。需要说明的是，Filename、Stream 和 Handlers 这 3 个参数中只能有 1 个存在，不能同时出现 2 个或 3 个，否则会引发 ValueError 异常。

关于 logging 模块日志格式字符串字段的介绍如表 8.3 所示。

表 8.3

字段/属性名称	使用格式	描述
asctime	%(asctime)s	日志事件发生的时间——可读时间，如 2023-01-07 16:49:45,896
created	%(created)f	日志事件发生的时间——时间戳
relativeCreated	%(relativeCreated)d	日志事件发生的时间与 logging 模块加载时间相比的相对毫秒数（目前还不知道用处）
msecs	%(msecs)d	日志事件发生时间的毫秒部分
levelname	%(levelname)s	日志信息的文字形式的日志等级（'DEBUG', 'INFO', 'WARNING', 'ERROR', 'CRITICAL'）
levelno	%(levelno)s	日志信息的数字形式的日志等级（10, 20, 30, 40, 50）
name	%(name)s	所使用的日志器名称，默认为'root'，因为默认使用的是 root logger
message	%(message)s	日志信息的文本内容，通过 msg % args 计算得到
pathname	%(pathname)s	调用日志记录方法的代码文件的全路径
filename	%(filename)s	pathname 的文件名部分，包含文件后缀
module	%(module)s	filename 的名称部分，不包含文件后缀
lineno	%(lineno)d	调用日志记录方法的代码所在的行号
funcName	%(funcName)s	调用日志记录方法的方法名
process	%(process)d	进程 ID
processName	%(processName)s	进程名称（Python 3 版本新增字段）
thread	%(thread)d	线程 ID
threadName	%(thread)s	线程名称

配置日志器的日志等级，代码如下：

```
import logging
logging.basicConfig(level=logging.DEBUG)
logging.log(logging.DEBUG,"I am a debug level log.")
logging.log(logging.INFO,"I am a info level log.")
logging.log(logging.WARNING,"I am a warning level log.")
logging.log(logging.ERROR,"I am a error level log.")
logging.log(logging.CRITICAL,"I am a critical level log.")
```

控制台上打印的内容如下：

```
DEBUG:root:I am a debug level log.
INFO:root:I am a info level log.
WARNING:root:I am a warning level log.
ERROR:root:I am a error level log.
CRITICAL:root:I am a critical level log.
```

以上所有等级的日志信息全部输出，说明配置已经生效。在配置了日志等级的基础上，再配置日志输出、日志文件和日志格式，代码如下：

```
import logging
LOG_FORMAT = "%(asctime)s - %(levelname)s - %(message)s"
logging.basicConfig(filename="log1.log",level=logging.DEBUG,format=LOG_FORMAT)
logging.log(logging.DEBUG,"I am a debug level log.")
logging.log(logging.INFO,"I am a info level log.")
logging.log(logging.WARNING,"I am a warning level log.")
logging.log(logging.ERROR,"I am a error level log.")
logging.log(logging.CRITICAL,"I am a critical level log.")
```

代码执行完毕，在当前目录下生成了一个日志文件"log1.log"，内容如下：

```
2019-01-06 19:54:42,093 - DEBUG - I am a debug level log.
2019-01-06 19:54:42,093 - INFO - I am a info level log.
2019-01-06 19:54:42,093 - WARNING - I am a warning level log.
2019-01-06 19:54:42,093 - ERROR - I am a error level log.
2019-01-06 19:54:42,093 - CRITICAL - I am a critical level log.
```

在以上配置的基础上，我们也可以加上日期/时间格式的配置，测试代码如下：

```
import logging
LOG_FORMAT = "%(asctime)s - %(levelname)s - %(message)s"
DATE_FORMAT = "%m/%d/%Y %H:%M:%S %p"
```

```
logging.basicConfig(filename='log6.log', level=logging.DEBUG, format=LOG_
FORMAT,datefmt=DATE_FORMAT)
    logging.log(logging.DEBUG,"I am a debug level log.")
    logging.log(logging.INFO,"I am a info level log.")
    logging.log(logging.WARNING,"I am a warning level log.")
    logging.log(logging.ERROR,"I am a error level log.")
    logging.log(logging.CRITICAL,"I am a critical level log.")
```

代码执行完毕,在当前目录下生成了一个日志文件"log6.log",内容如下:

```
2023-01-13 10:10:29,817 - DEBUG - I am a debug level log.
2023-01-13 10:10:29,817 - INFO - I am a info level log.
2023-01-13 10:10:29,817 - WARNING - I am a warning level log.
2023-01-13 10:10:29,817 - ERROR - I am a error level log.
2023-01-13 10:10:29,817 - CRITICAL - I am a critical level log.
```

以上是对 Python 日志的简单介绍,如果想对 Python 日志有更深入的了解,请参考 Python 官方文档。

下面定义一个 log 函数,目的是定义 logging 的 basicConfig 等信息。日志信息存放在当前目录下的 log-selenium.log 文件中,具体代码如下:

```
def log(str):
    logging.basicConfig(level=logging.INFO,
                format='%(asctime)s %(filename)s %(levelname)s %(message)s',
                datefmt='%a, %d %b %Y %H:%M:%S',
                filename='log-selenium.log',
                filemode='a')
    console = logging.StreamHandler()
    console.setLevel(logging.INFO)
    formatter = logging.Formatter('%(name)-12s: %(levelname)-8s %(message)s')
    console.setFormatter(formatter)
    logging.getLogger('').addHandler(console)
    logging.info(str)
```

8.2.3 初步实现数据驱动

通过以上对 Excel 文件操作和 Log 日志的学习,我们对数据和测试代码分离的思想有了初步认识,下面将以上知识应用到项目中。

functions.py 中新增了 read_excel 与 log 函数,代码如下:

```python
from selenium import webdriver
from selenium.webdriver.chrome.service import Service
from selenium.webdriver.common.by import By
import time
import xlrd
import logging
#这是新增函数,用于处理和获取 Excel 文件中的测试数据
def read_excel(filename,index):
    xls = xlrd.open_workbook(filename)
    sheet = xls.sheet_by_index(index)
    #print(sheet.nrows)
    #print(sheet.ncols)
    dic={}
    for j in range(sheet.ncols):
        data=[]
        for i in range(sheet.nrows):
          data.append(sheet.row_values(i)[j])
        dic[j]=data
    return dic
def log(str):
    logging.basicConfig(level=logging.INFO,
                format='%(asctime)s %(filename)s %(levelname)s %(message)s',
                datefmt='%a, %d %b %Y %H:%M:%S',
                filename='log-selenium.log',
                filemode='a')
    console = logging.StreamHandler()
    console.setLevel(logging.INFO)
    formatter = logging.Formatter('%(name)-12s: %(levelname)-8s %(message)s')
    console.setFormatter(formatter)
    logging.getLogger('').addHandler(console)
    logging.info(str)
```

对脚本代码文件 add_position.py 中的 add_position 函数进行优化,增加传入岗位名称和岗位编号的参数,代码更新如下:

```python
import time,os
from functions import return_driver,name,id,xpath
def add_position(postName,postcode,remark):
    driver = return_driver()
    driver.get("http://*******.site/index")
    time.sleep(15)
    #单击左侧导航栏->系统管理
    xpath('//*[@id="side-menu"]/li[3]/a/span[1]').click()
```

```python
time.sleep(5)
#单击左侧导航栏->系统管理->岗位管理
xpath('//*[@id="side-menu"]/li[3]/ul/li[5]/a').click()
time.sleep(2)
#单击"新增"超链接
#切换driver到iFrame，因为"新增"超链接位于iFrame内
driver.switch_to.frame("iframe6")
time.sleep(2)
xpath('//*[@id="toolbar"]/a[1]').click()
time.sleep(2)
#首先返回主页面，释放iFrame
driver.switch_to.default_content()
#再次进入新的iFrame
driver.switch_to.frame("layui-layer-iframe1")
#在添加岗位页面中，输入岗位名称
name('postName').send_keys(postName)
#在添加岗位页面中，输入岗位编码
name('postCode').send_keys(postCode)
#在添加岗位页面中，输入显示顺序
name('postSort').send_keys('1')
#在添加岗位页面中，输入备注
name('remark').send_keys(remark)
#首先返回主页面，释放iFrame
time.sleep(3)
driver.switch_to.default_content()
#在添加岗位页面中，单击"确定"按钮
xpath('//*[@id="layui-layer1"]/div[3]/a[1]').click()
time.sleep(5)
```

首先准备测试数据文件 positon_data.xls，其内容如图 8.30 所示，用于测试。

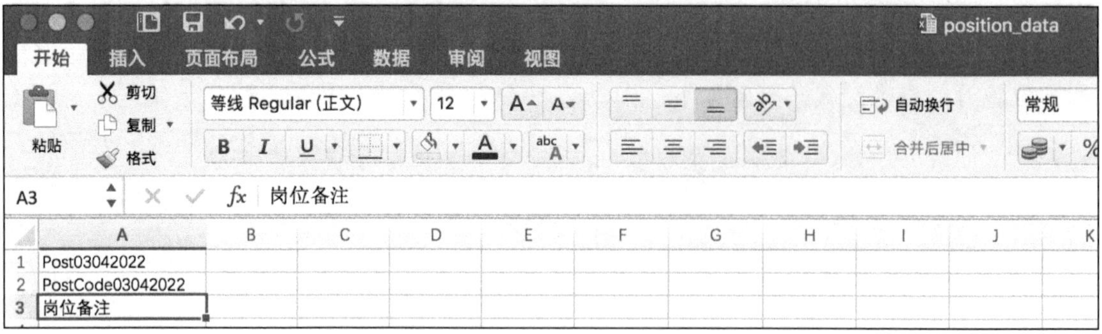

图 8.30

测试代码文件 test_system.py 具体内容如下,使用了上面 Excel 文件中的数据:

```
from login import login
from add_position import add_position
from functions import read_excel
from functions import log
#首先登录系统
log("Begin to log in system.")
login()
log("Log in system done.")
log("fetch position test data from Excel file")
dict1 = read_excel("position_data.xls",0)
log("Begin to add position in system.")
#测试添加岗位信息场景
add_position(dict1[0][0],dict1[0][1])
log("Add position done.")
```

以上测试代码可以成功执行,结果如图 8.31 所示,检查测试过程中的日志信息,请参考图 8.32。

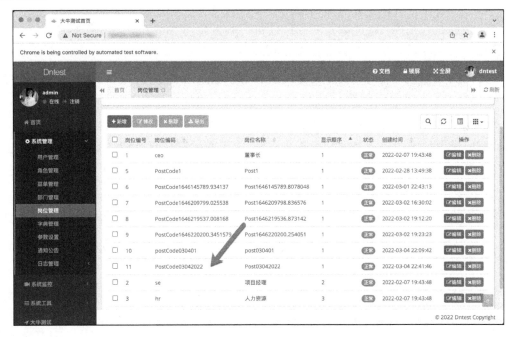

图 8.31

```
(selenium4.0-automation) (base) 192:Chapter10.2.3 jason118$ python test_system.py
root            : INFO     fetch position test data from Excel file
root            : INFO     Begin to add position in system.
root            : INFO     Begin to add position in system.
root            : INFO     Add position done.
```

图 8.32

8.3 数据驱动框架 DDT

8.3.1 单元测试

单元测试，百度百科上的解释是"对软件中的最小可测单元进行检查和验证"。对于单元的含义，在不同的编程语言中要根据实际情况进行判定，如在 C 语言中单元指一个函数，而在 Java 语言中单元可能指一个类。

从细节上看，单元测试是开发者编写的一小段代码，用于检验被测代码的一个很小、很明确的功能是否正确，通过单元测试可以将一些系统的 Bug 扼杀在初始阶段，这样付出的代价比在集成测试、系统测试阶段要小得多。

对面向对象编程语言 Python 来说，最小可测单元应该是类，在学习单元测试之前，需要先介绍一下 Python 类的相关知识。

类的定义和使用在面向对象的编程语言中很常见，是一种抽象的概念集合，表示一个共性产物，类中通常会定义属性和行为（方法）。下面举例说明类的结构并进行简单运用，有两点需要注意。

（1）在 Python 中，无论是函数还是类，定义范围时不使用其他语言常用的大括号"{}"而是使用缩进的方式，如下面例子中 info1 函数的函数体就只有一行"print("this is a pig")"。

（2）函数、类及判断语句声明部分结束后要以冒号"："结尾。请注意下例中函数和类声明行的结尾处：

```
#coding=utf-8
class Pig(object):
    def info1(self):
        print("this is a pig")
pig = Pig()
pig.info1()
```

这个例子中定义了一个类"Pig",Pig 类派生自 object。首先,定义类成员变量或方法,该类中定义了一个方法"info1"。然后,要实例化一个 pig 类。最后,实现调用方法 info1,输出字符串"this is a pig"。

以上只是一个简单的类的应用举例,让大家先对类有一个直观的认识。类的知识点有很多,由于篇幅限制,本书只会介绍比较常用和重要的知识点。

也可以将类理解为代码的另一种抽象,有些类似于函数,其目的之一是提供代码重用性和提升编码效率。在面向对象的编程语言中,类和实例都是特别重要的概念。示例代码如下,执行结果如图 8.33 所示。

```
#coding=utf-8
class Person(object):
    def __init__(self,name,age):
        self.name = name
        self.age = age
    def print_info(self):
        print('%s: %s' % (self.name,self.age))
p1 = Person('Jack',23)
print(p1.name)
print(p1.age)
```

```
daniu_8.3.1_2 ×
/usr/local/bin/python3 /Users/tim/Documents/pyse2023/chapter08/daniu_8_3_1/daniu_8.3.1_2.py
Jack
23

Process finished with exit code 0
```

图 8.33

以上代码主要介绍了一个常见的简单类的构造过程和实例化方法,类中主要包含构造方法__init__和方法 print_info。

一般来讲,类是描述一类事物的载体。在虚拟的代码世界里,如果要描述整个现实世界,就需要引入类这样的载体和方法。

举一个示例来说明一下类,以便我们更快地熟悉它的用法,代码如下:

```
#coding=utf-8
class animal:
#定义类的属性(动物年龄)
age =10
```

```python
    #Python的初始化方法
    def __init__(self,name):
        self.name=name
    #类的成员方法
    def eat(self):
        print("have something")
        print(self.name)
#定义子类bird，继承自父类animal
class bird(animal):
    #定义子类的初始化方法
    def __init__(self,name,color):
        #这里定义的方法是，如果子类中没有定义name属性，那么就继承父类的name属性，如
        #果子类中定义了name属性，那么就使用子类定义的属性
        super(bird,self).__init__(name)
        #定义属性color
        self.color=color
    #定义方法fly
    def fly(self):
        print(self.name)
        print(self.color)
#开始执行，有点类似Java的main方法，相当于程序的主入口，可以直接在命令行执行
if __name__=='__main__':
a = animal("xiaoli")
a.eat()
    #实例化对象的操作
    b =bird("xiaoniao","red")
    b.fly()
b.eat()
#以字典的形式打印对象b的属性值
    print(b.__dict__)
```

在上述代码的注释部分已经对代码行进行了必要的解释，代码的执行结果如图8.34所示。

```
xiaoniao
red
have something
xiaoniao
{'name': 'xiaoniao', 'color': 'red'}

Process finished with exit code 0
```

图 8.34

单元测试库（UnitTest）实现了我们在开发代码的过程中比较实际值和预期值等功能，使用起来很方便。UnitTest 作为一种单元测试框架，其思想来源于 JUnit，其与目前市面上主流的测试框架有很多相似之处。

下面简单介绍 UnitTest 工作流中的四大核心组件。

（1）Test Fixture 用于测试之前的准备工作，如数据清理工作、创建临时数据库、目录，以及开启某些服务进程等。

（2）Test Case 是最小测试单元，具有独立性。主要检测输出是否满足期望，其结果基于一系列特定的输入。UnitTest 提供了一个基类 TestCase，用来创建新的 Test Case。

（3）可以简单地将 Test Suite 理解为 Test Case 的集合，主要用于集成管理一起执行的测试用例。

（4）Test Runner 也是 UnitTest 的重要组件之一，主要用于协调测试的执行并将结果输出，给用户参考。

如图 8.35 所示是 UnitTest 中常用的断言。

Method	Checks that	New in
assertEqual(a, b)	a == b	
assertNotEqual(a, b)	a != b	
assertTrue(x)	bool(x) is True	
assertFalse(x)	bool(x) is False	
assertIs(a, b)	a is b	3.1
assertIsNot(a, b)	a is not b	3.1
assertIsNone(x)	x is None	3.1
assertIsNotNone(x)	x is not None	3.1
assertIn(a, b)	a in b	3.1
assertNotIn(a, b)	a not in b	3.1
assertIsInstance(a, b)	isinstance(a, b)	3.2
assertNotIsInstance(a, b)	not isinstance(a, b)	3.2

图 8.35

UnitTest 提供了丰富的工具集，可以创建和运行单元测试。

（1）所有的测试用例类都继承自基类 unittest.TestCase。Python 语法规定，父类要写在小括号内，如"XXXTest(unittest.TestCases)"。

（2）unittest.main 的作用是使一个单元测试模块变为可直接运行的测试脚本。

main 方法使用 TestLoader 类来搜索所有包含在当前测试类中以"test"开头的测试方法，并自动执行它们。执行方法的默认顺序是，根据 ASCII 码的顺序加载测试用例，数字与字母的顺序为 0~9、A~Z、a~z。因此 A 开头的方法会先执行，a 开头的方法会后执行。

（1）unittest.TestSuite，单元测试框架中的 TestSuite 类用于创建测试套件，其中最常用的一个方法是 addTest，该方法的功能是将测试用例添加到测试套件中。

（2）每个独立的单元脚本中的测试方法都应该以"test"字符串开头。

（3）assertEqual 方法的功能是验证实际执行结果是不是期望值。

（4）assertTrue 和 assertFalse 的功能是验证测试结果是否满足一定的条件。

（5）assertRaises 的功能是验证单元测试是否会抛出某一特定异常，如 TypeError。

（6）setUp 方法用于测试用例执行前的初始化工作，如在测试登录 Web 应用时，在 setUp 方法中执行实例化浏览器等操作。

（7）teardown 方法用于测试用例执行之后的善后操作，如关闭数据库连接、关闭浏览器等操作。

如下为单元测试练习代码：

```
import unittest
class add(unittest.TestCase):  #声明一个测试类
    def setUp(self):
        pass
    def test_01(self):
        self.assertEqual(2,2)
#test_01 方法的功能是判断 2 与 2 是否相等，预期结果为相等
    def test_02(self):
        self.assertEqual('selenium','appium')
#test_02 方法的功能是判断"selenium"字符串和"appium"字符串是否
#相等，预期结果为不相等
    def test_03(self):
        self.assertEqual('se','se')
#test_03 方法的功能是判断"se"和"se"是否是同样的字符串，预期结果是相同
    def tearDown(self):
        pass
if __name__ == '__main__':
    unittest.main()
```

代码的执行结果如图 8.36 所示。

```
! Tests failed: 1, passed: 2 of 3 tests – 4 ms
appium != selenium

Expected :selenium
Actual   :appium
 <Click to see difference>

Traceback (most recent call last):
  File "/Applications/PyCharm.app/Contents/helpers/pycharm/
    old(self, first, second, msg)
AssertionError: 'selenium' != 'appium'
- selenium
+ appium
```

图 8.36

下面我们用 UnitTest 运行一个 WebDriver 测试用例，业务场景如下。

（1）打开 Chrome 浏览器，打开 login.html 页面。

（2）清空用户名并输入"大牛测试"。

（3）检测页面上是否有"上传资料"字符串。

代码如下：

```python
#encoding = utf-8
import unittest
from selenium import webdriver
import time
class login(unittest.TestCase): #声明一个测试类
    def setUp(self):
        path = Service("/Users/tim/Downloads/chromedriver")
        self.driver = webdriver.Chrome(service=path)
        #打开登录页面
        self.driver.get('file:///Users/tim/Desktop/selenium/book/login.html')
    def test_01(self):
        self.driver.find_element(By.ID,"dntest").clear()
        self.driver.find_element(By.ID, "dntest").send_keys("大牛测试")
        self.assertIn("上传资料",self.driver.page_source)
    def tearDown(self):
        pass
if __name__ == '__main__':
    unittest.main()
```

在实际工作中,通常一个测试包含多个测试用例,这些测试用例可能来源于多个不同的模块。此时,利用自动化测试框架来批量执行,就可以省时省力,从而提高测试效率。

接下来,将具体介绍如何批量执行脚本。

(1) 创建一个项目"BatchRun"。

(2) 在项目上新建"Python Package",命令为"TestSuite"。

(3) 在 TestSuite 下新建文件夹"testset1"和"testset2"。

(4) 在文件夹 testset1 下,添加脚本文件"case01.py"和"case02.py"。

"case01.py"文件的代码如下:

```python
import unittest
import time
class Test1(unittest.TestCase):
    def setUp(self):
        print("开始执行脚本 01")
    def tearDown(self):
        time.sleep(3)
        print("脚本 01 执行结束!")
    def test_01(self):
        print("执行第 1 个用例!")
    def test_02(self):
        print("执行第 2 个脚本!")
    def test_03(self):
        print("执行第 3 个脚本!")
if __name__ == "__main__":
    unittest.main()
```

"case02.py"文件的代码如下:

```python
import unittest
import time
class Test2(unittest.TestCase):
    def setUp(self):
        print("开始执行脚本 02")
    def tearDown(self):
        time.sleep(3)
        print("脚本 02 执行结束!")
```

```
    def test_04(self):
        print("执行第 4 个用例！")
    def test_05(self):
        print("执行第 5 个脚本！")
    def test_06(self):
        print("执行第 6 个脚本！")
if __name__ == "__main__":
    unittest.main()
```

（5）在文件夹 testset2 下，添加脚本文件"case03.py"和"case04.py"。

"case03.py"文件的代码如下：

```
import unittest
import time
class Test1(unittest.TestCase):
    def setUp(self):
        print("开始执行脚本 03")
    def tearDown(self):
        time.sleep(3)
        print("脚本 03 执行结束！")
    def test_07(self):
        print("执行第 7 个用例！")
    def test_08(self):
        print("执行第 8 个脚本！")
    def test_09(self):
        print("执行第 9 个脚本！")
if __name__ == "__main__":
    unittest.main()
```

"case04.py"文件的代码如下：

```
import unittest
import time
class Test1(unittest.TestCase):
    def setUp(self):
        print("开始执行脚本 04")
    def tearDown(self):
        time.sleep(3)
        print("脚本 04 执行结束！")
    def test_10(self):
        print("执行第 10 个用例！")
```

```
    def test_11(self):
        print("执行第 11 个脚本!")
    def test_12(self):
        print("执行第 12 个脚本!")
if __name__ == "__main__":
    unittest.main()
```

最后,利用 UnitTest 的 discover 方法来实现测试脚本的批量执行。在文件夹 testsuite 下新建 Python 文件 "run_cases_inbatch.py",文件代码如下:

```
import unittest
import os
#查询测试用例路径
case_path = os.path.join(os.getcwd(),"testsuite")
def allcases():
    discover = unittest.defaultTestLoader.discover(case_path,pattern= "case*.py",top_level_dir=None)
    print(discover)
    return discover
if __name__ == "__main__":
    runner = unittest.TextTestRunner()
    runner.run(allcases())
```

通过执行以上代码,可以看到此次批量执行的脚本集合,discover 变量返回的字符串如下:

```
<unittest.suite.TestSuite tests=[<unittest.suite.TestSuite tests=[]>, <unittest.suite.TestSuite tests= [<unittest.suite.TestSuite tests=[<testset1.case01.Test1 testMethod=test_01>, <testset1.case01.Test1 testMethod=test_02>, <testset1.case01.Test1 testMethod=test_03>]>]>, <unittest.suite.TestSuite tests= [<unittest.suite.TestSuite tests=[<testset1.case06.Test2 testMethod=test_04>, <testset1.case06.Test2 testMethod=test_05>, <testset1.case06.Test2 testMethod=test_06>]>]>, <unittest.suite.TestSuite tests=[]>, <unittest.suite.TestSuite tests=[<unittest.suite.TestSuite tests=[<testset6.case06.Test1 testMethod= test_07>, <testset6.case06.Test1 testMethod=test_08>, <testset6.case06.Test1 testMethod=test_09>]>]>, <unittest.suite.TestSuite tests=[<unittest.suite.TestSuite tests=[<testset6.case06.Test1 testMethod=test_10>, <testset6.case06.Test1 testMethod=test_11>, <testset6.case06.Test1 testMethod=test_12>]>]>]>
```

项目整体结构如图 8.37 所示。

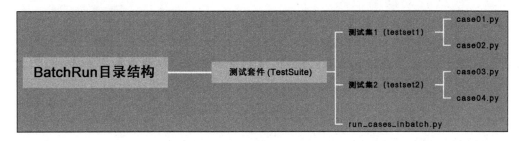

图 8.37

在批量执行测试脚本时，只需要执行 Python 文件"run_cases_inbatch.py"即可，执行结果如图 8.38 所示。

图 8.38

8.3.2 数据驱动框架应用

DDT 是"Data-Driven Tests"的缩写。UnitTest 没有自带数据驱动功能，在使用 UnitTest 的同时又想使用数据驱动框架 DDT，就需要安装 ddt 库，安装命令为"pip install ddt"。

使用方法如下。

（1）ddt.data，装饰测试方法，参数是元组、列表等。

（2）ddt.file_data，装饰测试方法，参数是文件名，测试数据保存在参数文件中。文件格式可以是 JSON 或者 YAML。

有一点需要注意，如果文件以".yml"结尾，ddt 会被当作 YAML 文件处理，其他文件都会被当作 JSON 文件处理。

（3）ddt.unpack，在 DDT 框架传递复杂的数据结构时使用。

下面演示一个简单的例子来说明 DDT 框架的用法，代码如下：

```
import ddt
import unittest
@ddt.ddt
class testse(unittest.TestCase):
    def setUp(self):
        pass
    @ddt.data(2,3)
    def test_01(self,t):
        print(t)
    def tearDown(self):
        pass
if __name__ == '__main__':
    unittest.main()
```

从代码分析可以知道，ddt 设置的参数列表是一个元组，并且这个元组的元素有 2 和 3。单元测试结果如图 8.39 所示，从结果来看，单元测试执行了两个测试用例，即测试方法 test_01 执行了两次。

```
[timdeMacBook-Pro:chapter08 tim$ python daniu_8.3.2_1.py
2
.3
.
----------------------------------------------------------------------
Ran 2 tests in 0.000s

OK
```

图 8.39

下面再举例说明 DDT 框架对 JSON 文件的用法，其中 JSON 文件内容为 "{"android": "appium", "web": "selenium", "interface": " python interface automation"}"，我们可以通过 JSON 文件来管理测试数据，具体代码如下：

```
import ddt
import unittest
@ddt.ddt
class test_se(unittest.TestCase):
    def setUp(self):
        pass
    @ddt.file_data("tt.json")   #将文件 tt.json 存放在当前文件夹内
    def test_01(self,tt):
        print(tt)
    def tearDown(self):
        pass
if __name__ == '__main__':
    unittest.main()
```

以上代码的执行结果如图 8.40 所示，JSON 文件键值对中的 value 全部被打印。

```
[timdeMacBook-Pro:chapter08 tim$ python daniu_8.3.2_2.py
appium
.selenium
.python interface automation
.
----------------------------------------------------------------------
Ran 3 tests in 0.000s

OK
```

图 8.40

在自动化测试结束后，往往需要查看执行结果，如何得到一份便于查看和管理的测试报告呢？在这里，笔者推荐 HTMLTestRunner 应用程序，它是 Python 标准库 UnitTest 模块的一个扩展，可以生成 HTML 的测试报告，而且界面十分友好。准备工作如下。

（1）从官网下载 HTMLTestRunner.py 文件。HTMLTestRunner 的下载界面如图 8.41 所示。需要注意的是，这里提供的 HTMLTestRunner 是 0.8.2 版本的，它的语法基于 Python 2。假如需要 Python 3 版本的 HTMLTestRunner，需要进行修改，网上有修改好的基于 Python 3 的 HTMLTestRunner，大家可以自行搜索并下载。

图 8.41

（2）将 HTMLTestRunner.py 文件复制到 Python 安装路径下的 lib 文件夹中。

（3）下面以 login.html 为例，展现 HTMLTestRunner 的用法。

测试代码如下：

```
#encoding = utf-8
import unittest
import HTMLTestRunner
from selenium import webdriver
from selenium.webdriver.chrome.service import Service
from selenium.webdriver.common.by import By
class login(unittest.TestCase): #声明一个测试类
```

```python
    def setUp(self):
        path = Service("/Users/tim/Downloads/chromedriver")
        self.driver = webdriver.Chrome(service=path)
        #打开登录页面
        self.driver.get('file:///Users/tim/Desktop/selenium/book/login.html')
    def test_01(self):
        self.driver.find_element(By.ID,"dntest").clear()
        self.driver.find_element(By.ID, "dntest").send_keys("大牛测试")
        self.assertIn("上传资料",self.driver.page_source)
    def tearDown(self):
        pass
if __name__ == '__main__':
    suite = unittest.TestSuite()
    suite.addTest(login("test_01"))
    #设置生成的报表HTML文件地址
    file_name = "/Users/tim/Documents/pyse2023/chapter08/daniureport.html"
    fp = open(file_name, 'wb')
    #设置报表页面的标题和报表总结描述内容
    runner = HTMLTestRunner.HTMLTestRunner
(stream=fp, title='Test_Report_Portal', description='Report_Description')
    runner.run(suite)
    fp.close()
    print("测试完成！")
```

最后，测试机器路径盘"/xx/chapter08/"，生成报表文件，报表内容截屏如图 8.42 所示，其中的"login"是单元测试脚本的类名。

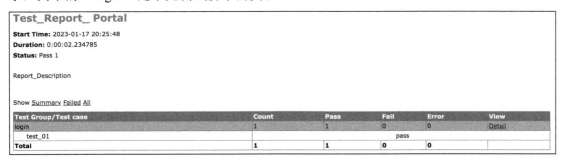

图 8.42

以上主要讲解了单元测试 UnitTest、HTMLTestRunner 和 DDT 框架的基本用法。将它们为测试所用，运用到实战中，才可以体现出价值。下面，笔者认为，是时候梳理一下本章的主要知识点了。

项目文件框架如图 8.43 所示。

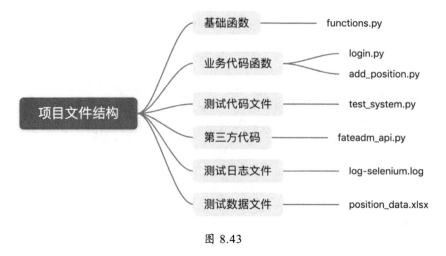

图 8.43

基础函数文件 functions.py 中的代码如下：

```
from selenium import webdriver
from selenium.webdriver.chrome.service import Service
from selenium.webdriver.common.by import By
import time
import xlrd
import logging
chrome_driver_server = Service("./chromedriver")
driver = webdriver.Chrome(service=chrome_driver_server)
'''
函数 return_driver 用于返回 driver 对象
'''
def return_driver():
    return driver
'''
函数 name 使用 name 定位方法来定位元素
'''
def name(locator):
    return driver.find_element(By.NAME,locator)
'''
函数 id 使用 id 定位方法来定位元素
'''
def id(locator):
    return driver.find_element(By.ID,locator)
```

```
'''
函数 xpath 使用 XPath 定位方法来定位元素
'''
def xpath(locator):
    return driver.find_element(By.XPATH,locator)
#这是新增函数，用于处理和获取 Excel 文件中的测试数据
def read_excel(filename,index):
    xls = xlrd.open_workbook(filename)
    sheet = xls.sheet_by_index(index)
    #print(sheet.nrows)
    #print(sheet.ncols)
    dic={}
    for j in range(sheet.ncols):
        data=[]
        for i in range(sheet.nrows):
            data.append(sheet.row_values(i)[j])
        dic[j]=data
    return dic
def log(str):
    logging.basicConfig(level=logging.INFO,
                format='%(asctime)s %(filename)s %(levelname)s %(message)s',
                datefmt='%a, %d %b %Y %H:%M:%S',
                filename='log-selenium.log',
                filemode='a')
    console = logging.StreamHandler()
    console.setLevel(logging.INFO)
    formatter = logging.Formatter('%(name)-12s: %(levelname)-8s %(message)s')
    console.setFormatter(formatter)
    logging.getLogger('').addHandler(console)
    logging.info(str)
```

业务代码文件 login.py 中的代码如下：

```
import time,os
import fateadm_api
from functions import return_driver, id, name,xpath
def login():
    driver = return_driver()
    driver.get("http://*******.site/login")
    filename = "capture.png"
    if os.path.exists(filename):
        os.remove(filename)
```

```
    ele1 = xpath('//*[@id="signupForm"]/div[1]/div[2]/a/img')
    ele1.screenshot(filename)
    name('username').clear()
    name('username').send_keys("admin")
    name('password').clear()
    name('password').send_keys("admin123")
    #如下调用了斐斐打码的代码,直接获取识别后的验证码值
    verification_code = str(fateadm_api.TestFunc())
    #在页面上输入验证码
    name('validateCode').send_keys(verification_code)
    #单击"登录"按钮
    id('btnSubmit').click()
    time.sleep(5)
```

业务代码文件 add_position.py 中的代码如下:

```
import time,os
from functions import return_driver,name,id,xpath
def add_position(postName,postCode):
    driver = return_driver()
    driver.get("http://*******.site/index")
    time.sleep(15)
    #单击左侧导航栏->系统管理
    xpath('//*[@id="side-menu"]/li[3]/a/span[1]').click()
    time.sleep(5)
    #单击左侧导航栏->系统管理->岗位管理
    xpath('//*[@id="side-menu"]/li[3]/ul/li[5]/a').click()
    time.sleep(2)
    #单击"新增"超链接
    #切换 driver 到 iFrame,因为"新增"超链接处于 iFrame 内
    driver.switch_to.frame("iframe6")
    time.sleep(2)
    xpath('//*[@id="toolbar"]/a[1]').click()
    time.sleep(2)
    #首先返回主页面,释放 iFrame
    driver.switch_to.default_content()
    #再次进入新的 iFrame
    driver.switch_to.frame("layui-layer-iframe1")
    #在添加岗位页面中,输入岗位名称
    name('postName').send_keys(postName)
    #在添加岗位页面中,输入岗位编码
    name('postCode').send_keys(postCode)
```

```
#在添加岗位页面中，输入显示顺序
name('postSort').send_keys('1')
#在添加岗位页面中，输入备注
name('remark').send_keys("remark text 1")
#首先返回主页面，释放 iFrame
time.sleep(3)
driver.switch_to.default_content()
#在添加岗位页面中，单击"确定"按钮
xpath('//*[@id="layui-layer1"]/div[3]/a[1]').click()
time.sleep(5)
```

测试代码文件 test_system.py 中的代码如下：

```
from login import login
from add_position import add_position
from functions import read_excel
from functions import log
import HTMLTestRunner
import unittest
import time
class system_add_position(unittest.TestCase):
    def test_add_position(self):
        log("Read Excel Files to get test data.")
        dict1 = read_excel("position_data.xlsx",0)
        log("Begin to log in system.")
        login()
        time.sleep(2)
        log("Log in system done.")
        log("Begin to add position in system.")
        #测试添加岗位信息场景
        add_position(dict1[0][0],dict1[0][1])
        log("Add position done.")
        time.sleep(2)
if __name__ == '__main__':
    suite = unittest.TestSuite()
    suite.addTest(system_add_position("test_add_position"))
    file_name = "report_add_position.html"
    fp = open(file_name,'wb')
    runner = HTMLTestRunner.HTMLTestRunner(stream=fp, title='Test_Report_Portal', description='Report_Description')
    #此步是为了设置报表页面的标题和报表总结描述内容
    runner.run(suite)
    fp.close()
```

测试数据 Excel 文件 position_data.xlsx 的内容如图 8.44 所示。

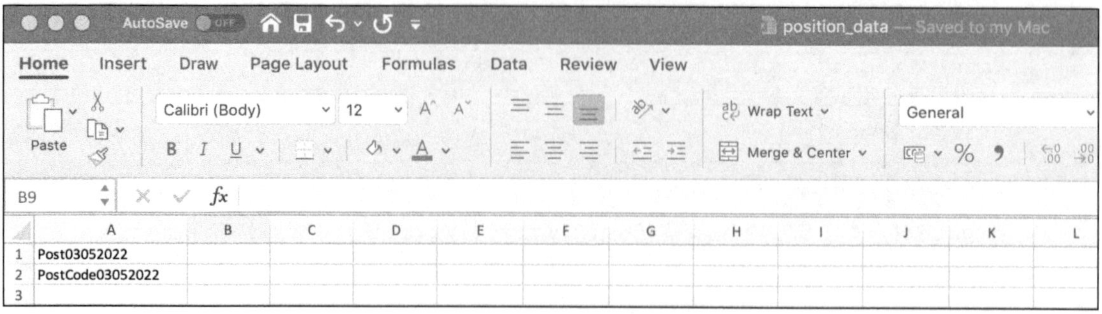

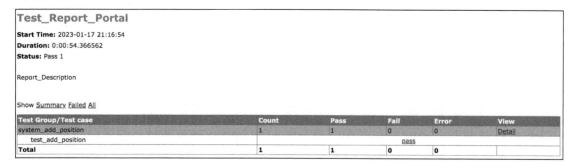

图 8.44

测试执行完成后，在当前目录下生成测试报告文件"report_add_position.html"，报告内容如图 8.45 所示。

图 8.45

最后检查测试日志文件 log-selenium.log，具体如图 8.46 所示。

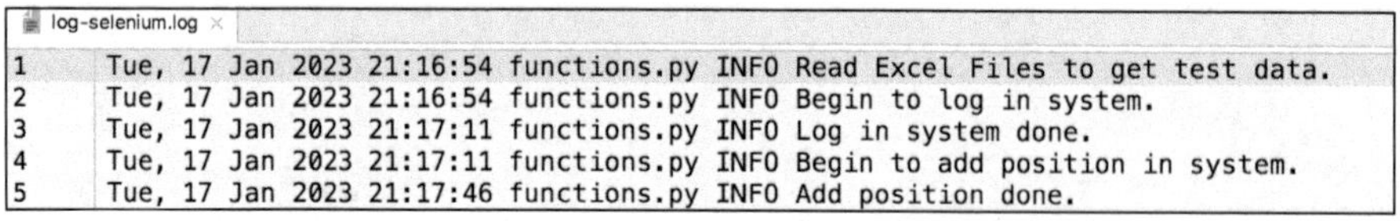

图 8.46

8.3.3 DDT+Excel 实现循环测试

在实际项目中，很多时候需要执行重复性测试而非一次性测试，大量的重复性测试才能体现出自动化测试的效率和价值。

接下来，以一个小案例来演示"如何运用 DDT 框架结合 Excel 文件类型的测试数据来实现自动化测试"，测试场景是批量用户登录。

项目文件结构如图 8.47 所示。

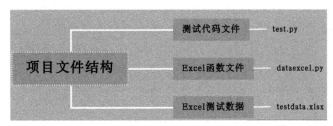

图 8.47

Excel 测试数据文件 testdata.xlsx 的内容如图 8.48 所示。

username	password
admin	admin123
daniu	daniu
daniutest	daniutest

图 8.48

Excel 函数文件 dataexcel.py 的内容如下，该文件的作用是提取测试数据并返回一个列表，而每个列表元素是一个字典对象：

```
#coding=utf-8
import xlrd
#封装读取 Excel 函数，函数返回的是一个字典
#需要修改一下参数名
#demo 数据返回[{'username': 'admin', 'passwd': 'admin123'},{'username': 'daniu', 'passwd': 'daniu'}, {'username': 'daniutest', 'passwd': 'daniutest'}]
def get_data(filename, sheetnum):
    path = 'testdata.xlsx'
    book_data = xlrd.open_workbook(path)
    book_sheet = book_data.sheet_by_index(1)   #打开文件中的第一个表
    rows_num = book_sheet.nrows  #行数
    rows0 = book_sheet.row_values(0)  #将第 1 行的各名称作为字典的键
    rows0_num = len(rows0)  #列数
    list = []
    for i in range(1, rows_num):
        rows_data = book_sheet.row_values(i)   #取每一行的值作为列表
```

```
        rows_dir = {}
        for y in range(0, rows0_num):  #将每一列的值与每一行对应起来
            rows_dir[rows0[y]] = rows_data[y]
        list.append(rows_dir)
    return list
if __name__ == '__main__':
    print(get_data('', 1))
```

测试代码文件 test.py 的内容如下，通过这个脚本可以实现循环测试，比较用户名字段与密码字段对应的字符串是否相同。如果不同，则测试失败，直到所有测试数据循环结束。

```
#coding=utf-8
from ddt import ddt ,data,file_data,unpack
from dataexcel import get_data
import unittest
from selenium import webdriver
excel=get_data('', 1)
@ddt
class test_se(unittest.TestCase):
    def setUp(self):
        path = Service("/Users/tim/Downloads/chromedriver")
        self.driver = webdriver.Chrome(service=path)
        self.driver.get('http://testdao.site/login')
#对字典进行操作
    @data(*excel)
    def test_01(self,dic):
        self.driver.find_element(By.NAME,'username').send_keys(dic.get ('username'))
        self.driver.find_element(By.NAME,'password').send_keys(dic.get('passwd'))
        print(dic)
        self.assertEqual(dic.get('username'),dic.get('passwd'))
    def tearDown(self):
        pass
if __name__ == '__main__':
    unittest.main()
```

在命令行窗口中，切换到脚本所在的目录并执行代码，命令为"python test.py"，执行结果如图 8.49 所示。

```
⊘ Tests failed: 3 of 3 tests – 7 s 962 ms
     self.driver.find_element(By.NAME,'password').send_keys(dic.get('passwd'))
  File "/Library/Frameworks/Python.framework/Versions/3.11/lib/python3.11/site
     Command.SEND_KEYS_TO_ELEMENT, {"text": "".join(keys_to_typing(value)), "va
  File "/Library/Frameworks/Python.framework/Versions/3.11/lib/python3.11/site
     for i in range(len(val)):
TypeError: object of type 'NoneType' has no len()


Ran 3 tests in 7.965s

FAILED (errors=3)

Process finished with exit code 1
```

图 8.49

从图 8.49 可以看出，3 次测试方法的执行都是失败的，因为期望值与实际值不相等，因此断言取值为比较用户名与密码是否相等。

第 9 章 Page Object 设计模式

本章采用类对所有模块进行重构,每一个页面对象与方法都封装在一个类中,所有模块都用结构化方式展示,进而形成可落地的测试框架。

9.1 什么是 Page Object

Page Object 的缩写是 PO,中文翻译为"页面对象模式"。它是一种设计模式,目的是创建 Web UI 对象库,即涉及的每一个页面都被定义为一个单独的类。类中应该包含页面元素对象和处理这些元素对象所需的方法等。方法的命名需要遵守一定的规则,能清楚、正确地表明方法的作用。

PO 设计模式的优点如下。

（1）PO 提供了页面元素操作和业务流程相分离的模式，使测试的代码结构比之前更清晰，可读性更强。

（2）更方便地重用对象和方法。

（3）对象库可以通过集成不同的工具类来达到不同的测试目的。如集成 UnitTest 可以做单元测试自动化/功能测试，同时也可以集成 JBehave/Cucumber 等来做验收测试。

（4）方便自动化测试的维护和更新，若某个页面的元素需要变更，仅需要修改封装好的页面元素类，而不用修改测试类。

PO 的核心思想是分层，把同属于一个页面的元素都封装在一个页面类中。如对于登录页面，设计 3 个不同的类来体现这种分层思想，达到 PO 的目的。即以页面为单位，将某一个页面中的元素控件等全部提炼出来并封装成相应的方法，形成可以被调用的对象。

9.2　Page Object 实战

在 9.1 节中，已经简单地介绍了 PO，大家对 PO 模式的原理和特性有了初步的了解。但是要真正掌握 PO 的精髓，至少还需要掌握两点：一是对被测系统的功能点要有充分的认识；二是要落地项目，才能体现自动化测试的意义。

本节选择第 8 章的项目实战案例，笔者将一步一步地顺着项目的脉络带大家循序渐进地熟悉 PO 的思想。

PO 项目的框架结构图，如图 9.1 所示，在图中可以清晰地看到分层情况，每层的具体功能会在本章后面进行详细讲解。

图 9.1 是对 PO 项目总体框架的介绍，其中的核心层是 PO 层（Page Object 层），围绕着 PO 层，又新增了其他层级。采用结构化模型可以使自动化测试越来越清晰，也会降低后期维护成本，提高项目自动化的程度。

Base 层封装了项目需要的基础方法，如元素 click 操作、send keys 操作，调用 JavaScript 脚本的方法和其他一些与基本浏览器相关的操作等。

Base 层的一些方法调用了 Common 层的模块方法，需要先熟悉 Common 层代码结构。

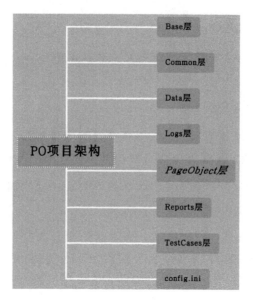

图 9.1

9.2.1　Common 层代码分析

Common 层主要包含处理 Excel 文件的方法，获取项目路径、测试系统 URL 的信息和日志功能的实现方法。接下来，将对具体实现细节和重要的知识点进行讲解。

（1）获取项目路径、测试系统 URL 的方法，文件名为"function.py"，代码如下：

```python
import os,configparser
#获取项目路径
def project_path():
    return os.path.split(os.path.realpath(__file__))[0].split('C')[0]
#返回config.ini 文件中 testUrl
def config_url():
    config = configparser.ConfigParser()
    config.read(project_path() + "config.ini")
    return config.get('testUrl', 'url')
if __name__ == '__main__':
    print("项目路径为："+project_path())
    print("被测系统 URL 为："+config_url())
```

这部分代码运用了 os 和 configparse 两个 Python 模块。os 模块主要用于对操作系统级别的目录、文件夹进行操作和对文件进行操作（读取、写入等）；而 configparse 模块的主要功能是读取配置文件。方法 project_path 的功能是获取项目的当前目录。方法 config_url 的功能是获取被测系统的网址信息，其中涉及自动化项目的配置文件 config.ini，内容如图 9.2 所示，其中的 url 值就是项目中被测系统的网址（大牛测试系统地址）。

```
config.ini
1  [testUrl]
2  url = http://localhost/login
3  [productUrl]
4  url = http://localhost/login
```

图 9.2

以上 function.py 脚本执行结果如图 9.3 所示，证明方法功能运行正常。

```
function
/usr/local/bin/python3 /Users/tim/Documents/pyse2023/chapter09/Common/function.py
项目路径为：/Users/tim/Documents/pyse2023/chapter09/
被测系统URL为：http://localhost/login

Process finished with exit code 0
```

图 9.3

（2）创建日志类，便于在项目中添加日志信息，文件名为"log.py"。类名为 Framelog，主要方法为 log，作用是返回一个 logger 对象。其中比较重要的几点是，设置日志文件的文件名命名方式，设置日志文件的存放路径，设置日志输出内容格式和布局信息，这是日志的核心功能，具体代码如下：

```python
import logging,time
from Common.function import project_path
class FrameLog():
    def __init__(self, logger=None):
    #创建一个logger
        self.logger = logging.getLogger(logger)
        self.logger.setLevel(logging.DEBUG)
    #创建一个handler，用于写入日志文件
        self.log_time = time.strftime("%Y_%m_%d_")
        #路径需要修改
        self.log_path = project_path() + "/Logs/"
        self.log_name = self.log_path + self.log_time + 'log.log'
```

```
        print(self.log_name)
        fh = logging.FileHandler(self.log_name, 'a', encoding='utf-8')
        fh.setLevel(logging.INFO)
        #定义handler的输出格式
        formatter = logging.Formatter('[%(asctime)s] %(filename)s->%(funcName)s line:%(lineno)d [%(levelname)s]%(message)s')
        fh.setFormatter(formatter)
        self.logger.addHandler(fh)
        #在记录日志后移除句柄
        self.logger.removeHandler(fh)
        #关闭打开的文件
        fh.close()
    def log(self):
        return self.logger
if __name__ == '__main__':
    lo = FrameLog()
    log = lo.log()
    log.error("error")
    log.debug("debug")
    log.info("info")
    log.critical("严重")
```

从测试结果看，日志存储在项目主目录的"/Logs/2023_01_04_log.log"文件中，如图9.4所示。

图9.4

日志文件内容如图9.5所示。

```
  2023_01_14_log.log
1 [2023-01-14 10:16:02,844] log.py-><module> line:36 [ERROR]error
2 [2023-01-14 10:16:02,844] log.py-><module> line:38 [INFO]info
3 [2023-01-14 10:16:02,845] log.py-><module> line:39 [CRITICAL]严重
4 [2023-01-14 10:22:41,879] log.py-><module> line:36 [ERROR]error
5 [2023-01-14 10:22:41,880] log.py-><module> line:38 [INFO]info
6 [2023-01-14 10:22:41,880] log.py-><module> line:39 [CRITICAL]严重
```

图9.5

（3）处理测试数据文件（Excel 文件），代码文件名为"exceldata.py"，代码如下：

```python
import xlrd,os
#读取Excel文件的操作，将所有的数据存放在字典中
#filename为文件名
#index为Excel sheet工作簿索引
def read_excel(filename,index):
    xls = xlrd.open_workbook(filename)
    sheet = xls.sheet_by_index(index)
    print(sheet.nrows)
    print(sheet.ncols)
    dic={}
    for j in range(sheet.ncols):
        data=[]
        for i in range(sheet.nrows):
            data.append(sheet.row_values(i)[j])
        dic[j]=data
    return dic
if __name__ == '__main__':
    #读取Excel文件的操作，返回字典
    data = read_excel(project_path()+"/Data/testdata.xlsx",0)
    print(data)
    print(data.get(1))
```

在上面这段代码中，主要方法为 read_excel，即读取存储在 Excel 文件中的测试数据，方法最后会返回字典对象。"print(data)"打印的是字典对象，字典对象的细节为"{0: ['岗位名称', '岗位编号'], 1: ['Post0305012022', 'PostCode0305012022'], 2: ['Post0305022022', 'PostCode0305022022']}"。测试代码"print(data.get(1))"返回的是列表对象，列表对象的细节为"['Post0305012022', 'PostCode0305012022']"。

读者也可以继续对字典对象进行处理，项目中用到的测试数据如图 9.6 所示。

	A	B	C
1	岗位名称	Post0305012022	Post0305022022
2	岗位编号	PostCode0305012022	PostCode0305022022
3			

图 9.6

以上对 Excel 文件的操作使用单个方法封装，也可以用类的方式直接封装，代码如下：

```python
import xlrd
from chapter09.Common.function import project_path
class ExcelUtil:
    def __init__(self,excelpath=None,sheet=None):
        self.excelpath = excelpath
        if sheet ==None:
            sheet = "Sheet1"
        self.workbook = xlrd.open_workbook(self.excelpath)
        self.sheet = self.workbook.sheet_by_name(sheet)
    #行数
    def getrows(self):
        rows = self.sheet.nrows
        if rows >= 1:
            return rows
        return None
    #列数
    def getcols(self):
        cols = self.sheet.ncols
        if cols >= 1:
            return cols
        return None
    #获取单元格数据
    def getcell(self,row,col):
        if self.getrows() > row:
            data = self.sheet.cell(row,col).value
            return data
        return None
    #一次性读取全部数据并存放在字典中
    def ExcelDic(self):
        dic = {}
        for r in range(self.getrows()):
            list = []
            for c in range(self.getcols()):
                list.append(self.sheet.row_values(r)[c])
            dic[r] =list
        return dic
if __name__ == "__main__":
    path =project_path()+"Data/"+"testdata.xlsx"
    ex = ExcelUtil(path,"Sheet2")
```

```
dic = ex.ExcelDic()
print(dic[0])
print(dic[0][1])
```

以文件 testdata.xlsx 中的 Sheet2 工作表为例,运行代码后的执行结果如图 9.7 所示。

```
excelutil
/usr/local/bin/python3 /Users/tim/Documents/pyse2023/chapter09/Common/excelutil.py
['用户名', 'admin']
admin

Process finished with exit code 0
```

图 9.7

9.2.2　Base 层代码分析

Base 层代码在项目中涉及底层操作,如对 click、send_keys 及 clear 等事件的封装,可以提高代码的重用性。文件 base.py 中主要包含定位元素方法 findele,方法 findele 返回的结果是定位元素的语句,该方法用参数*args 接收任意多个非关键字参数,参数的类型是元组。代码如下:

```
from Common.log import FrameLog
from selenium.webdriver import ActionChains
from selenium.webdriver.common.by import By
from selenium.webdriver.support.select import Select
from common.Log import framelog
from selenium.webdriver.support.ui import WebDriverWait
from selenium.webdriver.support import expected_conditions as ex
#对 Base 层代码进行优化
class Base():
    def __init__(self,driver):
        self.driver = driver
        self.log =FrameLog().log()
    #单星号参数代表接收任意多个非关键字参数
    def findele(self,*args):
        try:
            print(args)
            self.log.info("通过"+args[0]+"定位,元素是"+args[1])
            return   WebDriverWait(self.driver,20).until(ex.visibility_of(self.driver.find_element(*args)))
        except:
```

```python
            #在页面上没有定位到相应的元素
            self.log.error("定位元素失败!")
#对元素执行click操作
def click(self,*args):
    self.findele(*args).click()
#输入值
def sendkey(self,*args,value):
    self.findele(*args).send_keys(value)
#调用js方法
def js(self,str):
    self.driver.execute_script(str)
def url(self):
    return self.driver.current_url
#后退
def back(self):
    self.driver.back()
#前进
def forword(self):
    self.driver.forward()
#退出
def quit(self):
    self.driver.quit()
#切换到主页方法,在需要切换到主页面时调用该方法
def go_content(self):
    self.driver.switch_to.default_content()
#窗口最大化
def maxwin(self):
    self.driver.maximize_window()
#全屏
def fulwin(self):
    self.driver.fullscreen_window()
#最小化
def minwin(self):
    self.driver.minimize_window()
#切换窗口,针对两个窗口进行切换
def newtab(self):
    h = self.driver.window_handles
    self.driver.switch_to.window(h[-1])
#获取文本
def gettext(self,*args):
    return self.findele(*args).text
#下拉列表框,在其中选择某一个元素
```

```python
    def select(self,*args,num):
        se = self.findele(*args)
        Select(se).select_by_index(num)
    #alert
    def alert(self):
        self.driver.switch_to.alert.accept()
    #获取属性
    def getattr(self,*args,attr):
        return self.findele(*args).get_attribute(attr)
    #单击鼠标右键
    def contextclick(self,*args):
        ActionChains(self.driver).context_click(self.findele(*args)).perform()
    #单击鼠标左键
    def clickhold(self,*args):
        ActionChains(self.driver).click_and_hold(self.findele(*args)).perform()
    #双击
    def doubleclick(self,*args):
        ActionChains(self.driver).double_click(self.findele(*args)).perform()
    #悬停
    def movetoele(self,*args):
        ActionChains(self.driver).move_to_element(self.findele(*args)).perform()
    #拖拉操作
    def drag(self,*args):
        sour = self.driver.find_element(By.CSS_SELECTOR, args[0])
        slide = self.driver.find_element(By.CSS_SELECTOR, args[1])
        ActionChains(self.driver).drag_and_drop_by_offset(sour, slide.size['width'], -slide.size["height"]).perform()
```

编写好基础代码之后需要进行单元测试，以 findele、click 方法为例，测试代码如下：

```python
from chapter09.Base.base import Base
from selenium.webdriver.chrome.service import Service
from selenium.webdriver.common.by import By
from selenium import webdriver
path = Service("/Users/tim/Downloads/chromedriver")
driver = webdriver.Chrome(service=path)
#打开登录页面
driver.get('file:///Users/tim/Desktop/selenium/book/login.html')
b = Base(driver)
#调用 findele 方法
b.findele(By.ID,"dntest").send_keys('大牛测试')
#调用 click 方法
b.click(By.ID,"loginbtn")
```

注意,以上使用显式等待 WebDriverWait 查找元素的方法进行封装,同时封装了其他常用方法,如双击、单击鼠标左键等方法。

Base 层还包含另一个文件 base_unit.py,该文件的目的是抽离单元测试中的 setUp 和 tearDown 方法,代码如下:

```
#conding:uft-8
import unittest
from Common.function import config_url
from selenium import webdriver
#抽离单元测试中的 setUp 与 tearDown 方法
class UnitBase(unittest.TestCase):
    @classmethod
    def setUpClass(cls):
        cls.driver = webdriver.Chrome()
        cls.driver.get(config_url())
        cls.driver.maximize_window()
    def tearDownClass(cls):
        cls.driver.quit()
```

9.2.3 PageObject 层代码分析

PageObject 是核心层,该层不但涉及代码技术,还涉及对项目业务的分析,进而对相关的页面进行分析。在业务分析方面,首先分析要进行 PO 的页面,其次对每个范围内的页面进行详细分析(如自动化需要用到的元素和相关的操作方法,以及页面之间的关联情况等)。

在本书中,笔者将继续使用第 8 章的案例作为 PO 实战案例。主要测试场景包括大牛测试系统登录页面、后台首页和添加岗位页面。

1. 系统登录页面

大牛测试系统登录页面的代码(文件名为"login_page.py")如下:

```
from Chapter09.Base.base_update import Base
from selenium.webdriver.common.by import By
import time,os
from Chapter09.PageObject import fateadm_api
class LoginPage(Base):
```

```python
#以下为系统登录页面的 4 个元素定位语句
def login_user(self):
    return self.findele(By.NAME,"username")
def login_password(self):
    return self.findele(By.NAME,"password")
def login_validateCode(self):
    return self.findele(By.NAME,"validateCode")
def login_validateImage(self):
    return self.findele(By.XPATH,'//*[@id="signupForm"]/div[1]/div[2]/a/img')
def login_button(self):
    return self.findele(By.ID,"btnSubmit")
#以下为登录系统的方法
def login_system(self,username,password):
    filename = "capture.png"
    if os.path.exists(filename):
        os.remove(filename)
    self.login_validateImage().screenshot(filename)
    self.login_user().clear()
    self.login_user().send_keys(username)
    self.login_password().clear()
    self.login_password().send_keys(password)
    verification_code = str(fateadm_api.TestFunc())
    self.login_validateCode().send_keys(verification_code)
    self.login_button().click()
    return self.url()
```

以上代码类 LoginPage 继承自基类 Base。另外，封装了页面基础元素，如方法 login_user 是用户名元素的定位语句，其调用了 Base 类下面的 findele 方法，findele 方法的传入参数分别是 By.NAME 和 "username" 字符串。

方法 login_system 封装了输入用户名、密码、验证码、登录操作，return 可以返回当前页面的 URL 地址。

2. 后台首页

在后台首页上，主要涉及页面元素系统管理和岗位管理，页面文件名为 "index_page.py"，代码如下：

```python
from Chapter09.Base.base_update import Base
from selenium.webdriver.common.by import By
```

```
import time
class IndexPage(Base):
    def index_sysadmin(self):
        return self.findele(By.XPATH,'//*[@id="side-menu"]/li[3]/a/span[1]')
    def index_postadmin(self):
        return self.findele(By.XPATH,'//*[@id="side-menu"]/li[3]/ul/li[5]/a')
    def position_menu(slef):
        self.index_sysadmin().click()
        self.index_postadmin().click()
```

以上代码页面类 IndexPage 继承自基类 Base。方法 index_sysadmin 和 index_postadmin 用于定位元素信息，position_menu 是封装后的方法。

3. 添加岗位页面

成功进入添加岗位页面后，此时的需求是新增岗位信息。涉及如下 6 个元素，"新增"超链接、岗位名称、岗位编码、显示顺序、备注、"确定"按钮。页面文件名为"add_position_page.py"，代码如下：

```
from Chapter09.Base.base_update import Base
from selenium.webdriver.common.by import By
import time
class AddPositionPage(Base):
    #定位"新增"HTML 超链接元素
    def add_position_insert(self):
        return self.findele(By.XPATH,'//*[@id="toolbar"]/a[1]')
    #定位岗位名称输入框
    def add_position_name(self):
        return self.findele(By.NAME,'postName')
    #定位岗位编码输入框
    def add_position_code(self):
        return self.findele(By.NAME,'postCode')
    #定位显示顺序输入框
    def add_position_order(self):
        return self.findele(By.NAME,'postSort')
    #定位备注多行文本输入框
    def add_position_remark(self):
        return self.findele(By.NAME,'remark')
    #定位"确定"按钮
```

```python
    def add_position_confirm(self):
        return self.findele(By.XPATH,'//*[@id="layui-layer1"]/div[3]/a[1]')
    #切换到iFrame,在单击"新增"岗位信息超链接前调用
    def go_frame_1(self):
        self.driver.switch_to.frame("iframe6")
    #切换到主页方法,在需要切换到主页面时调用该方法
    def go_content(self):
        self.driver.switch_to.default_content()
    #切换到iFrame,在输入岗位名称前调用
    def go_frame_2(self):
        self.driver.switch_to.frame("layui-layer-iframe1")
    #添加岗位信息方法,输入参数为岗位名称和岗位编码
    def add_post(self,postName,postcode,order,remark):
        self.go_frame_1()
        self.add_position_insert().click()
        self.go_content()
        self.go_frame_2()
        self.add_position_code().send_keys(postCode)
        self.add_position_order().send_keys(order)
        self.add_position_remark().send_keys(remark)
        self.go_content()
        self.add_position_confirm().click()
```

以上代码类 AddPositionPage 继承自基类 Base，方法 add_position_insert、add_position_name、add_position_code、add_position_order、add_position_remark 和 add_position_confirm 是用于返回页面相关元素的定位语句。方法 add_post 实现了添加岗位信息，需要输入参数岗位名称和岗位编码。

9.2.4　TestCases 层代码分析

TestCases 层的作用是管理测试用例和执行测试，相当于测试的总入口。在项目中，该层可以存放测试套件和测试用例的相关代码。

如上所述，首先定义测试总入口文件的代码，代码文件名为"suite.py"，代码如下：

```
#coding=utf-8
import unittest
import HTMLTestRunner
```

```
import time
from Common.function import project_path
if __name__ == '__main__':
    test_dir= project_path() + "TestCases"
    tests=unittest.defaultTestLoader.discover(test_dir,
                                              pattern ='test*.py',
                                              top_level_dir=None)
    now = time.strftime("%Y-%m-%d-%H_%M_%S",time.localtime(time.time()))
    filepath= project_path() + "/Reports/" + now + '.html'
    fp=open(filepath,'wb')
    #定义测试报告的标题与描述
    runner = HTMLTestRunner.HTMLTestRunner(stream=fp,title=u'自动化测试报告',description=u'测试报告')
    runner.run(tests)
    fp.close()
```

代码分析：在测试管理过程中，引入了 HTMLTestRunner 模块，该模块是 UnitTest 模块的一个扩展。为了更方便地生成测试报告，需要与单元测试 UnitTest 模块结合在一起使用。

首先，在代码中定义项目测试用例（测试代码）的存放地址，即项目主目录下的 TestCases 文件夹。然后，定义测试用例规则，在规则中定义测试目录、测试文件的模式等。最后，定义测试报告的文件路径和文件名等的规则，测试报告存放地址为项目主目录下的 Reports 文件夹。

9.2.5 Data 层分析

项目中的 Data 目录用于存放测试数据模块，该模块用于维护测试数据，如本项目中的测试数据文件 testdata.xls 等，这时对 Data 层的管理和维护就显得特别重要。从笔者的经验出发，Data 层通常有以下几点需要注意。

（1）数据文件类型的选型。在本项目中，关于数据的存储，选用的是 Excel 文件。读者要根据项目的实际需要灵活选择，笔者在第 8 章中讲解过不同的文件类型。举例说明，在测试场景中，有时需要测试上传功能，在测试数据中需要清楚定义数据字段等头部信息，而且上传文件的信息可能比较大（如包含 10 万行数据），此时用 CSV 文件就比用 Excel

文件优势明显。原因是，CSV 保存的是文本文件，而 Excel 保存的是二进制文件，从软件的可操作性或易用性来说，Excel 文件比不上 CSV 文件，而在大数据量场景下，Excel 文件没有 CSV 文件高效。

（2）数据的维护要有条理性。关联性比较强的测试数据可存放在同一个工作表中，便于维护。

9.2.6　Logs 层分析

Logs 层数据主要存放在项目运行过程中产生的日志文件中。日志文件记录了每次测试执行过程中的详细信息，便于分析、定位测试过程中的异常。框架的日志在一定程度上反映了框架是否运行在正确的轨道上。关于日志层面的维护，笔者的一些建议如下。

（1）日志等级或频度要设置得合理。在执行测试用例时，需要对要打印的日志进行筛选，对于一些重点操作最好打印日志。

（2）建议给框架中添加一些系统级别的日志。如对系统资源进行监控的日志，这样可以便于对一些非框架问题进行定位。笔者曾经遇到过由磁盘空间问题导致的框架异常。

（3）日志的管理，如定期清理日志。

本项目的框架采用自带的 log 模块，也可以使用第三方模块 loguru，使用模块前应先安装"pip install logugu"，测试代码如下：

```
from loguru import logger
logger.debug("This's a log message")
logger.info("this is a log info")
logger.error("this is a log error")
```

运行代码后的执行结果如图 9.8 所示。

```
/usr/local/bin/python3 /Users/tim/Documents/pyse2023/chapter09/Common/logu.py
2023-01-14 11:34:52.329 | DEBUG    | __main__:<module>:2 - This's a log message
2023-01-14 11:34:52.329 | INFO     | __main__:<module>:3 - this is a log info
2023-01-14 11:34:52.329 | ERROR    | __main__:<module>:4 - this is a log error

Process finished with exit code 0
```

图 9.8

封装 loguru 方法，生成日志数据并存放于 logs 目录下，文件格式如图 9.9 所示，代码如下：

```python
from chapter09.Common.function import project_path
from loguru import logger
def log():
    path =project_path()+"Logs/"
    log_file =path+'log_{time: YYYY_MM_DD}.log'
    #移除控制台
    #logger.remove(handler_id=None)
    logger.add(log_file, rotation='200kb', compression='zip')
    return logger
if __name__ == '__main__':
    print(project_path())
    log = log()
    log.info("大牛测试 info")
    log.debug("大牛测试 debug")
    log.error("大牛测试 error")
```

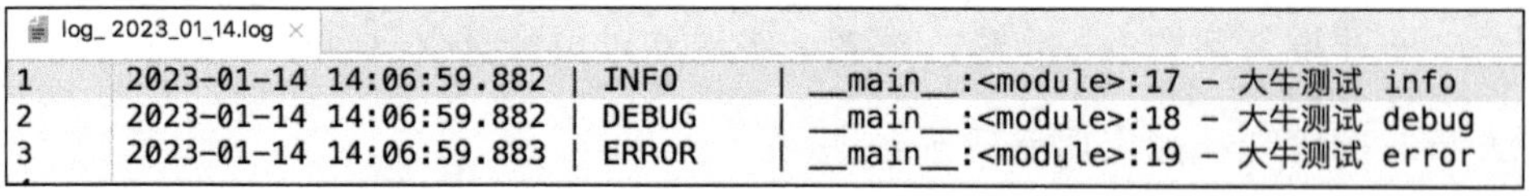

```
log_ 2023_01_14.log
1  2023-01-14 14:06:59.882 | INFO  | __main__:<module>:17 - 大牛测试 info
2  2023-01-14 14:06:59.882 | DEBUG | __main__:<module>:18 - 大牛测试 debug
3  2023-01-14 14:06:59.883 | ERROR | __main__:<module>:19 - 大牛测试 error
```

图 9.9

9.2.7 Reports 层分析

Reports 层数据主要存放在项目执行过程中产生的测试报告文件中，测试报告是对测试结果的总结。Reports 层的报表可能不只是一个测试报告文件那么简单，还需要探究报告所呈现内容的准确性、完整性和持续性等。而且，后续报表功能的迭代不能影响之前的测试结果或报告。

如果自动化测试项目的周期比较长，建议对测试结果进行数据存储，如将数据存储在 MongoDB 或者 MySQL 数据库中等。

9.2.8 其他分析

除了以上各层，在项目主目录下还需要维护一个配置文件（文件名是"config.ini"），该文件包含了整个项目的配置项，在本章的 Common 层内容中已经做过介绍。如果后续还需要添加其他项目，直接在该文件中进行维护即可。建议将驱动 driver 存放在项目根目录下，如图 9.10 所示。

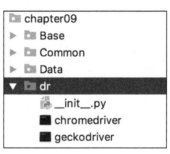

图 9.10

9.2.9 执行 Page Object 项目

通过以上对 PO 项目的介绍，相信大家已经对 PO 落地有了一定的了解。接下来将执行项目并查看测试报告和日志是否达到了预期。

执行代码时，需要执行 TestCases 下的 suite.py，这是管理用例的入口，代码如下：

```
# _*_ coding:utf-8 _*_
import unittest
import HTMLTestRunner
import time
from Common.function import project_path
if __name__ == '__main__':
    test_dir= project_path() + "TestCases"
    tests=unittest.defaultTestLoader.discover(test_dir,
                                              pattern ='login*.py',
                                              top_level_dir=None)
    now = time.strftime("%Y-%m-%d-%H_%M_%S",time.localtime(time.time()))
    filepath= project_path() + "/Reports/" + 'report1.html'
```

```
        fp=open(filepath,'wb')
        #定义测试报告的标题与描述
        runner = HTMLTestRunner.HTMLTestRunner(stream=fp,title=u'自动化测试报告',
description=u'测试报告')
        runner.run(tests)
        fp.close()
```

对以上代码进行分析可以发现，此次执行测试用例的脚本是以"login"开头的.py 文件，测试报告存放在项目的默认路径"xx/Reports"下。

在编写测试脚本前需要先熟悉 setUp、tearDown、setUpClass 和 tearDownClass 方法的使用场景。在多个测试用例中使用 setUp 方法，浏览器会被初始化多次。各方法的区别如下。

- setUp 方法：在测试方法运行开始前执行。

- tearDown 方法：在测试方法运行结束后执行。

- setUpClass 方法：在测试方法运行开始前执行，需要用@classmethod 装饰器装饰，测试过程中只执行一次

- tearDownClass 方法：在测试方法运行结束后执行，需要用@classmethod 装饰器装饰，测试过程中只执行一次

setUp 测试代码如下：

```
import unittest
class add(unittest.TestCase):  #声明一个测试类
    def setUp(self):
        print("测试开始")
    def test_01(self):
        print('test01')
        self.assertEqual(2,2)
    def test_02(self):
        print('test02')
        self.assertEqual('大牛测试','大牛测试')
    def tearDown(self):
        print("测试结束")
```

```
if __name__ == '__main__':
    unittest.main()
```

代码执行效果如图 9.11 所示，在每个测试方法前后都打印"测试开始""测试结束"。

```
timdeMacBook-Pro:TestCases tim$ python test_setup.py
测试开始
test01
测试结束
.测试开始
test02
测试结束
.
----------------------------------------------------------------------
Ran 2 tests in 0.000s

OK
```

图 9.11

setUpClass 测试代码如下：

```
import unittest
class addcla(unittest.TestCase):    #声明一个测试类
    @classmethod
    def setUpClass(cls):
        print("测试开始")
    def test_01(self):
        print('test01')
        self.assertEqual(2,2)
    def test_02(self):
        print('test02')
        self.assertEqual('大牛测试','大牛测试')
    @classmethod
    def tearDownClass(cls):
        print("测试结束")
if __name__ == '__main__':
    unittest.main()
```

代码执行效果如图 9.12 所示，"测试开始"与"测试结束"只打印一次。

```
timdeMacBook-Pro:TestCases tim$ python test_setupclass.py
测试开始
test01
.test02
.测试结束
----------------------------------------------------------------
Ran 2 tests in 0.000s

OK
```

图 9.12

为简单起见,这里只编写了一个测试文件"login_test.py",测试代码如下:

```
#_*_coding:utf-8_*_
import time,unittest,HTMLTestRunner
from Chapter09.Common.function import project_path
from Chapter09.Common.excel_data import read_excel
from Chapter09.Common.function import config_url
from Chapter09.PageObject.login_page import LoginPage
from selenium import webdriver
from selenium.webdriver.chrome.service import Service
class login_test(unittest.TestCase):
@classmethod
    def setUpClass(cls):
        cls.data = read_excel(project_path() + "/Data/testdata.xls", 1)
        chrome_driver_server = Service("./chromedriver")
        cls.driver = webdriver.Chrome(service=chrome_driver_server)
        cls.driver.get(config_url())
        cls.driver.maximize_window()
    def test_01(self):
        login = LoginPage(self.driver)
        res = login.login_system(self.data.get(1)[0],self.data.get(1)[1])
        self.assertIn("",res)
    def test_02(self):
        index = IndexPage(self.driver)
        res = index.position_menu()
        self.assertIn("", res)
    def test_03(self):
        add = AddPositionPage(self.driver)
        res = add.add_post(self.data.get(1)[0],self.data.get(1)[1])
        self.assertIn("", "")
@classmethod
```

```
    def tearDownClass(cls):
        cls.driver.quit()
```

测试结果文件如图 9.13 所示，而测试报告的内容如图 9.14 所示，测试脚本运行通过。如果测试失败，则可在报告中或者测试日志中寻找错误的详细信息，进而分析原因。

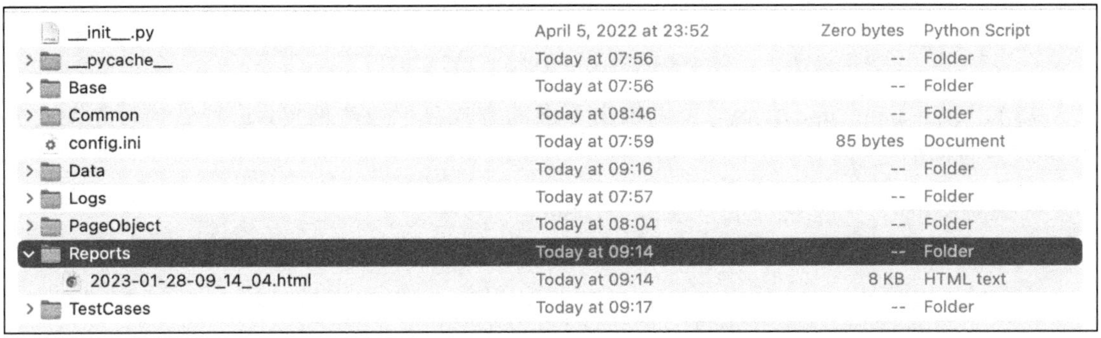

图 9.13

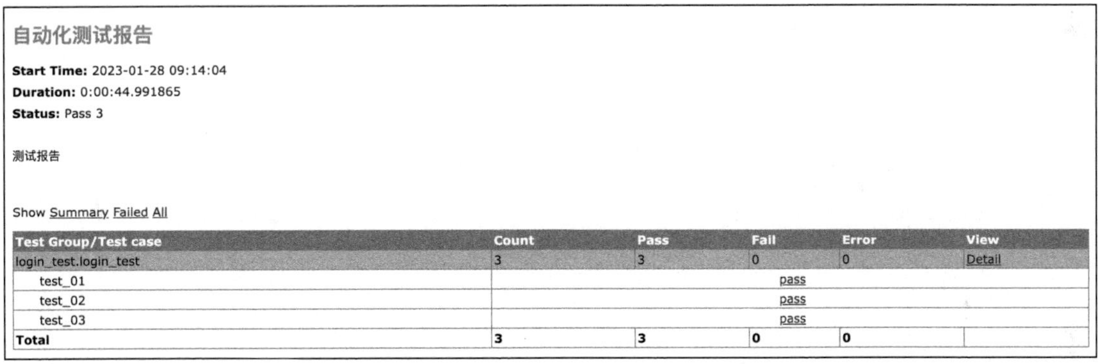

自动化测试报告

Start Time: 2023-01-28 09:14:04
Duration: 0:00:44.991865
Status: Pass 3

测试报告

Show Summary Failed All

Test Group/Test case	Count	Pass	Fail	Error	View
login_test.login_test	3	3	0	0	Detail
test_01	pass				
test_02	pass				
test_03	pass				
Total	3	3	0	0	

图 9.14

继续优化 Common 层，在 5.17 节学习过 By.ID="id"，下面以 login_page 为例进行优化。

在 Base.py 中添加如下方法：

```
    def findelement(self, locator):
        try:
            self.log.info("通过" + locator[0] + "定位,元素是" + locator[1])
            return WebDriverWait(self.driver, 20).until(ex.presence_of_element_
located(locator))
        except:
            #在页面上没有定位到相应的元素
```

```
            self.log.error("定位元素失败！")
#对输入框方法 send_keys 进行封装
def send(self,locator,value):
    self.findelement(locator).send_keys(value)
#对 click 方法进行封装
def clickbtn(self,locator):
    self.findelement(locator).click()
```

添加 login_page_update.py，可以根据下面的方式自行修改框架，这种结构较之前的更简单。修改后的文件代码如下：

```
from chapter09.Base.base import Base
from selenium.webdriver.common.by import By
import time,os
from chapter09.PageObject import fateadm_api
class LoginPage(Base):
    #用户名
    l_user = "name","username"
    #密码
    l_passwd = "name","password"
    #验证码
    l_validate = "name","validateCode"
    #验证码图片
    l_image = "xpath",'//*[@id="signupForm"]/div[1]/div[2]/a/img'
    #"登录"按钮
    l_login = "id","btnSubmit"
    #以下为登录系统的方法
    def login_system(self,username,password):
        self.send(self.l_user,username)
        self.send(self.l_passwd,password)
        self.clickbtn(self.l_login)
```

第 10 章 pytest 框架实战

前面章节使用 UnitTest 作为单元测试框架,本章将引入另一主流单元测试框架 pytest 以及 Allure 报告来对项目进行重构。

10.1 pytest 与 Allure

pytest 是一款功能比较完善的基于 Python 的测试框架。pytest 是由 UnitTest 单元测试框架衍生出来的新框架,使用起来更方便。使用 UnitTest 编写的测试用例可以在 pytest 单元测试框架中使用,反之则不可以。

10.1.1 pytest 的安装

使用 pip 命令安装 pytest：

```
pip install pytest
```

安装完成之后，查看安装结果，如图 10.1 所示，证明 pytest 安装完毕，安装的是当前的最新版本 7.2.1（截至本书完稿时）。

```
[timdeMacBook-Pro:chapter10 tim$ pip list|grep pytest
pytest                  7.2.1
```

图 10.1

10.1.2 简单测试案例介绍

新建测试文件"test_instance01.py"，内容如下。方法 return_name 会返回名字"jason"，而方法 test_return_name 用于测试方法 return_name 返回的内容是否等于"jason"。

```
def return_name():
    return "jason"
def test_return_name():
    assert return_name() == "jason"
```

在当前文件夹下，运行命令"#pytest + 文件名"，运行结果如图 10.2 所示。pytest 测试文件以"test"开头或结尾，测试方法需要以"test"开头。

```
[timdeMacBook-Pro:chapter10 tim$ python -m pytest test_instance01.py
============================= test session starts ==============================
platform darwin -- Python 3.11.1, pytest-7.2.1, pluggy-1.0.0
rootdir: /Users/tim/Documents/pyse2023/chapter10
plugins: assume-2.4.3, html-3.2.0, metadata-2.0.4
collected 1 item

test_instance01.py .                                                     [100%]

============================== 1 passed in 0.01s ===============================
```

图 10.2

10.1.3 引入类来管理测试方法

如果要执行很多测试方法且这些测试方法之间有关联,为了更好地管理测试方法,可以引入类。下面以具体案例进行阐述。新建文件"test_instance02.py",内容如下,在测试类 TestPersonalInfo 中定义了 3 个方法和 1 个测试方法:

```
class TestPersonalInfo:
    def return_name(self):
        return "jason"
    def return_age(self):
        return 25
    def return_job(self):
        return "engineer"
    def test_1(self):
        assert self.return_name() == "jason"
        assert self.return_age() == 25
        assert self.return_job() == "engineer"
```

执行代码之后的测试结果如图 10.3 所示,说明测试通过。

```
[timdeMacBook-Pro:chapter10 tim$ python -m pytest test_instance02.py
=========================== test session starts ===========================
platform darwin -- Python 3.11.1, pytest-7.2.1, pluggy-1.0.0
rootdir: /Users/tim/Documents/pyse2023/chapter10
plugins: assume-2.4.3, html-3.2.0, metadata-2.0.4
collected 1 item

test_instance02.py .                                               [100%]

============================ 1 passed in 0.01s ============================
```

图 10.3

以上案例的测试结果展现的内容过于简单,pytest 还提供了命令行参数。在相同的目录下,执行命令"pytest –v",测试结果如图 10.4 所示。测试结果中增加了测试文件中的测试方法名或测试类名:测试方法名。并且,表示测试通过与否的格式也不太一样,之前测试通过时会显示一个绿色圆点,现在显示的是绿色的"PASSED"字样。

```
[timdeMacBook-Pro:chapter10 tim$ python -m pytest -v test_instance02.py
=============================== test session starts ===============================
platform darwin -- Python 3.11.1, pytest-7.2.1, pluggy-1.0.0 -- /Library/Framewo
rks/Python.framework/Versions/3.11/bin/python3
cachedir: .pytest_cache
metadata: {'Python': '3.11.1', 'Platform': 'macOS-10.13.6-x86_64-i386-64bit', 'P
ackages': {'pytest': '7.2.1', 'pluggy': '1.0.0'}, 'Plugins': {'assume': '2.4.3',
 'html': '3.2.0', 'metadata': '2.0.4'}}
rootdir: /Users/tim/Documents/pyse2023/chapter10
plugins: assume-2.4.3, html-3.2.0, metadata-2.0.4
collected 1 item

test_instance02.py::TestPersonalInfo::test_1 PASSED                         [100%]

=============================== 1 passed in 0.01s ================================
```

图 10.4

pytest 还包含如下一些常用参数，对测试也比较有帮助。

- -x，表示在测试过程中，如果有一个测试用例执行失败，则退出整个测试，常用于调试测试代码。

- -k，用于对测试用例的筛查，如只运行包含"search"字样的测试用例，命令为"pytest -k search test_case.py"。

- -s，可以显示测试代码中的 print 语句的输出。

当只需要执行某一个测试方法时，可以使用 pytest 中的指定测试用例功能来实现。如命令"pytest test_instance02.py::TestPersonalInfo::test_1"的作用是执行 test_instance02.py 测试文件中的测试类 TestPersonalInfo 的测试方法 test_1。

10.1.4　setup 和 teardown 方法应用

setup 和 teardown 方法可用于完善测试过程。setup 方法可以在执行测试方法之前执行，而 teardown 方法可以在执行测试方法之后执行。这两个函数是分级别的，不同级别的作用范围和用途不太一样，其中比较常用的级别是方法级别和类级别。

首先来了解下方法级别，新建文件"test_instance03.py"，代码如下：

```
class TestPersonalInfo:
    def return_name(self):
        return "jason"
    def return_age(self):
        return 25
    def return_job(self):
```

```
            return "engineer"
    def setup(self):
        print("this is the setup function.")
    def teardown(self):
        print("this is the tear down function.")
    def test_1(self):
        assert self.return_name() == "jason"
        assert self.return_age() == 25
        assert self.return_job() == "engineer"
    def test_2(self):
        assert self.return_name() == "jason"
        assert self.return_age() == 25
        assert self.return_job() == "engineer"
```

运行命令"pytest -sv test_instance03.py"来执行测试，测试结果如图 10.5 所示，方法级别的 setup/teardown 会在执行每个测试方法时被执行。

```
[timdeMacBook-Pro:chapter10 tim$ python -m pytest -sv test_instance03.py
========================= test session starts =========================
platform darwin -- Python 3.11.1, pytest-7.2.1, pluggy-1.0.0 -- /Library/Framewo
rks/Python.framework/Versions/3.11/bin/python3
cachedir: .pytest_cache
metadata: {'Python': '3.11.1', 'Platform': 'macOS-10.13.6-x86_64-i386-64bit', 'P
ackages': {'pytest': '7.2.1', 'pluggy': '1.0.0'}, 'Plugins': {'assume': '2.4.3',
 'html': '3.2.0', 'metadata': '2.0.4'}}
rootdir: /Users/tim/Documents/pyse2023/chapter10
plugins: assume-2.4.3, html-3.2.0, metadata-2.0.4
collected 2 items

test_instance03.py::TestPersonalInfo::test_1 this is the setup function.
PASSEDthis is the tear down function.

test_instance03.py::TestPersonalInfo::test_2 this is the setup function.
PASSEDthis is the tear down function.
```

图 10.5

再来看类级别的 setup_class 和 teardown_class，新建文件 test_instance04.py，代码如下：

```
class TestPersonalInfo:
    def return_name(self):
        return "jason"
    def return_age(self):
        return 25
    def return_job(self):
        return "engineer"
    def setup_class(self):
```

```python
        print("this is the setup class.")
    def teardown_class(self):
        print("this is the tear down class.")
    def test_1(self):
        assert self.return_name() == "jason"
        assert self.return_age() == 25
        assert self.return_job() == "engineer"
    def test_2(self):
        assert self.return_name() == "jason"
        assert self.return_age() == 25
        assert self.return_job() == "engineer"
```

运行命令"pytest -sv test_instance04.py"来执行测试，测试结果如图 10.6 所示，执行后 setup_class 和 teardown_class 方法在整个测试过程中只运行了 1 次。

```
timdeMacBook-Pro:chapter10 tim$ python -m pytest -sv test_instance04.py
================= test session starts =================
platform darwin -- Python 3.11.1, pytest-7.2.1, pluggy-1.0.0 -- /Library/Framewo
rks/Python.framework/Versions/3.11/bin/python3
cachedir: .pytest_cache
metadata: {'Python': '3.11.1', 'Platform': 'macOS-10.13.6-x86_64-i386-64bit', 'P
ackages': {'pytest': '7.2.1', 'pluggy': '1.0.0'}, 'Plugins': {'assume': '2.4.3',
 'html': '3.2.0', 'metadata': '2.0.4'}}
rootdir: /Users/tim/Documents/pyse2023/chapter10
plugins: assume-2.4.3, html-3.2.0, metadata-2.0.4
collected 2 items

test_instance04.py::TestPersonalInfo::test_1 this is the setup class.
PASSED
test_instance04.py::TestPersonalInfo::test_2 PASSEDthis is the tear down class.

================= 2 passed in 0.01s =================
```

图 10.6

10.1.5　fixtures 功能应用

fixtures 是 pytest 的重要功能之一，一般被 fixtures 标记的方法可以预先执行，且该方法可以将 Python 对象以变量的方式传给测试方法。下面以一个案例来讲解 fixtures 的用法。新建文件"test_instance05.py"，内容如下，总结如下。

- 使用 fixtures 功能时，需要引入 pytest 模块。
- 案例中定义了两个类，在创建 Car 实例时，可以设置加油量多少，以升为单位。

- 定义被 pytest.fixture 标记的方法 "get_oil_litre_for_cars"。
- 创建测试方法 "test_oil_litres"，并且将上面定义的方法以变量的方式传入测试方法。这样做的目的是判断每一个创建的 Car 实例的加油量是不是大于 10，如果大于 10，则测试通过，反之测试失败。代码如下：

```
import pytest
class Refill_Oil:
    def __init__(self,oil_litre):
        self.oil_litre = oil_litre
class Car(Refill_Oil):
    def get_oil_litre(self):
        return self.oil_litre
@pytest.fixture
def get_oil_litre_for_cars():
    return [Car(11),Car(12),Car(20)]
def test_oil_litres(get_oil_litre_for_cars):
    car_list = get_oil_litre_for_cars
    for car in car_list:
        assert car.get_oil_litre() > 10
```

执行测试命令 "pytest -sv test_instance05.py"，测试结果如图 10.7 所示，表示测试通过。

```
timdeMacBook-Pro:chapter10 tim$ python -m pytest -sv test_instance05.py
============================= test session starts =============================
platform darwin -- Python 3.11.1, pytest-7.2.1, pluggy-1.0.0 -- /Library/Framewo
rks/Python.framework/Versions/3.11/bin/python3
cachedir: .pytest_cache
metadata: {'Python': '3.11.1', 'Platform': 'macOS-10.13.6-x86_64-i386-64bit', 'P
ackages': {'pytest': '7.2.1', 'pluggy': '1.0.0'}, 'Plugins': {'assume': '2.4.3',
 'html': '3.2.0', 'metadata': '2.0.4'}}
rootdir: /Users/tim/Documents/pyse2023/chapter10
plugins: assume-2.4.3, html-3.2.0, metadata-2.0.4
collected 1 item

test_instance05.py::test_oil_litres PASSED

============================== 1 passed in 0.01s ==============================
```

图 10.7

10.1.6　pytest 如何做参数化

在测试过程中，很可能需要基于同一个测试方法进行多轮测试，但是每轮测试的输入参数可能是不一样的。此时需要考虑测试数据的管理和应用。pytest 单元测试框架提供了参数化的方式，下面以一个案例来说明其用法，新建文件"test_instance06.py"，代码如下：

```
import pytest
@pytest.mark.parametrize("param1",[10,2,3,4])
def test_method1(param1):
    assert param1 > 1
```

执行命令"pytest -sv test_instance06.py"，测试结果如图 10.8 所示，表明测试方法 test_method1 执行了 4 次。

```
timdeMacBook-Pro:chapter10 tim$ python -m pytest -sv test_instance06.py
============================= test session starts ==============================
platform darwin -- Python 3.11.1, pytest-7.2.1, pluggy-1.0.0 -- /Library/Framewo
rks/Python.framework/Versions/3.11/bin/python3
cachedir: .pytest_cache
metadata: {'Python': '3.11.1', 'Platform': 'macOS-10.13.6-x86_64-i386-64bit', 'P
ackages': {'pytest': '7.2.1', 'pluggy': '1.0.0'}, 'Plugins': {'assume': '2.4.3',
 'html': '3.2.0', 'metadata': '2.0.4'}}
rootdir: /Users/tim/Documents/pyse2023/chapter10
plugins: assume-2.4.3, html-3.2.0, metadata-2.0.4
collected 4 items

test_instance06.py::test_method1[10] PASSED
test_instance06.py::test_method1[2] PASSED
test_instance06.py::test_method1[3] PASSED
test_instance06.py::test_method1[4] PASSED

============================== 4 passed in 0.01s ===============================
```

图 10.8

10.1.7 conftest 应用

conftest 文件是 pytest 特有的配置文件，可以实现数据、参数、方法、函数的共享。fixture 中的 scope 参数可以控制 fixture 的作用范围，共有 function、scope、module 和 session 4 种作用范围，具体介绍如下：

- function：每一个函数或方法都会被调用。新建文件"conftest.py"，代码如下：

```
import pytest
@pytest.fixture(scope="function")
def auto():
    print("daniutest")
```

新建文件"test05.py"，代码如下：

```
import pytest
@pytest.mark.usefixtures("auto")
class Test1:
```

```
    def test_01(self):
        print("test01")
    def test_02(self):
        print("test02")
if __name__=="__main__":
    pytest.main(["-s","test05.py"])
```

代码执行结果如图 10.9 所示，每个测试方法都会执行。

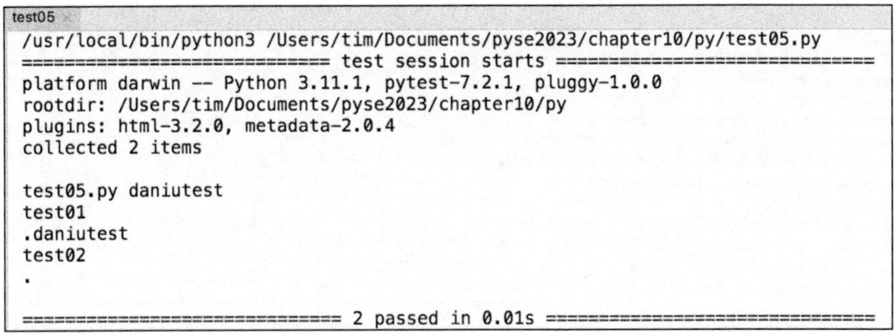

```
test05
/usr/local/bin/python3 /Users/tim/Documents/pyse2023/chapter10/py/test05.py
============================ test session starts =============================
platform darwin -- Python 3.11.1, pytest-7.2.1, pluggy-1.0.0
rootdir: /Users/tim/Documents/pyse2023/chapter10/py
plugins: html-3.2.0, metadata-2.0.4
collected 2 items

test05.py daniutest
test01
.daniutest
test02
.
============================== 2 passed in 0.01s =============================
```

图 10.9

- class：每一个类被调用一次。

把 conftest.py 文件中的 scope 修改为 "class"：

```
import pytest
@pytest.fixture(scope="class")
def auto():
    print("daniutest")
```

代码执行结果如图 10.10 所示，整个测试过程仅执行 1 次 auto 方法。

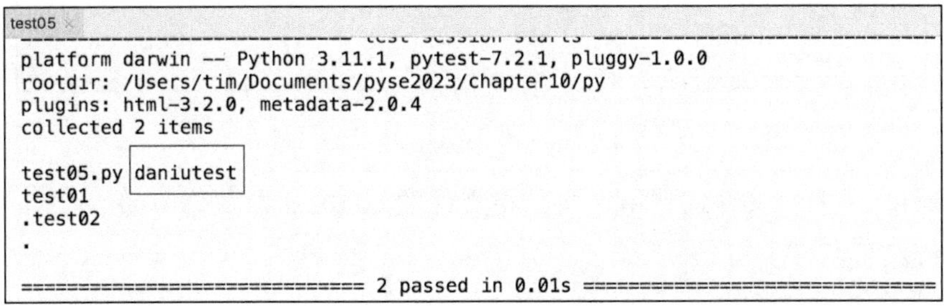

```
test05
============================ test session starts =============================
platform darwin -- Python 3.11.1, pytest-7.2.1, pluggy-1.0.0
rootdir: /Users/tim/Documents/pyse2023/chapter10/py
plugins: html-3.2.0, metadata-2.0.4
collected 2 items

test05.py daniutest
test01
.test02
.
============================== 2 passed in 0.01s =============================
```

图 10.10

- module：每一个 Python 文件被调用一次。

```
import pytest
@pytest.fixture(scope="module")
def auto():
    print("daniutest")
```

新增文件"test06.py"，代码如下，代码执行结果如图 10.11 所示。

```
import pytest
@pytest.mark.usefixtures("auto")
class Test1:
    def test_03(self):
        print("test03")
    def test_04(self):
        print("test04")
if __name__=="__main__":
    pytest.main(["-s","test06.py"])
```

```
[timdeMacBook-Pro:pyscope tim$ python -m pytest -sv test*.py
============================= test session starts =============================
platform darwin -- Python 3.11.1, pytest-7.2.1, pluggy-1.0.0 -- /Library/Framewo
rks/Python.framework/Versions/3.11/bin/python3
cachedir: .pytest_cache
metadata: {'Python': '3.11.1', 'Platform': 'macOS-10.13.6-x86_64-i386-64bit', 'P
ackages': {'pytest': '7.2.1', 'pluggy': '1.0.0'}, 'Plugins': {'assume': '2.4.3',
 'html': '3.2.0', 'metadata': '2.0.4'}}
rootdir: /Users/tim/Documents/pyse2023/chapter10/pyscope
plugins: assume-2.4.3, html-3.2.0, metadata-2.0.4
collected 4 items

test05.py::Test1::test_01 daniutest
test01
PASSED
test05.py::Test1::test_02 test02
PASSED
test06.py::Test1::test_03 daniutest
test03
PASSED
test06.py::Test1::test_04 test04
PASSED

============================== 4 passed in 0.01s ==============================
```

图 10.11

- session：多个文件仅调用一次，将文件的 conftest 参数修改为"session"：

```
import pytest
@pytest.fixture(scope="session")
def auto():
    print("daniutest")
```

代码执行结果如图 10.12 所示。

图 10.12

10.1.8 运行 Selenium

以大牛测试系统登录为例，用 conftest 配置文件重构代码，目录结构如图 10.13 所示，conftest.py 文件代码如下：

```
import pytest
from selenium import webdriver
from selenium.webdriver.chrome.service import Service
@pytest.fixture(scope="module")
def dr():
    path = Service("/Users/tim/Downloads/chromedriver")
    driver = webdriver.Chrome(service=path)
    return driver
```

图 10.13

test_login.py 文件代码如下，test 方法中增加了参数 dr：

```
import pytest,time
```

```python
from selenium.webdriver.common.by import By
class Test:
    @pytest.fixture(scope="function",autouse=True)
    def open_daniu(self,dr):
        dr.get("http://*******.site/login")
    def test_01(self,dr):
        dr.find_element(By.NAME, "username").clear()
        dr.find_element(By.NAME, "username").send_keys("admin")
        dr.find_element(By.NAME, "password").clear()
        dr.find_element(By.NAME, "password").send_keys("admin123")
        assert "login" in dr.current_url
        time.sleep(5)
if __name__ =="__main__":
    pytest.main(["-s","test_login.py"])
```

10.1.9 使用 pytest 生成测试报告

pytest 框架会生成 HTML 测试报告,下面以 test_instance06.py 文件为例生成测试报告。首先需要安装软件包 pytest-html,安装命令为"pip install pytest-html"。

执行命令"pytest --html=report1.html -sv test_instance06.py",测试结果如图 10.14 所示,从最后一行可以看到生成的测试报告文件目录。

```
timdeMacBook-Pro:chapter10 tim$ python -m pytest --html=report1.html -sv test_in
stance06.py
============================ test session starts ============================
platform darwin -- Python 3.11.1, pytest-7.2.1, pluggy-1.0.0 -- /Library/Framewo
rks/Python.framework/Versions/3.11/bin/python3
cachedir: .pytest_cache
metadata: {'Python': '3.11.1', 'Platform': 'macOS-10.13.6-x86_64-i386-64bit', 'P
ackages': {'pytest': '7.2.1', 'pluggy': '1.0.0'}, 'Plugins': {'assume': '2.4.3',
 'html': '3.2.0', 'metadata': '2.0.4'}}
rootdir: /Users/tim/Documents/pyse2023/chapter10
plugins: assume-2.4.3, html-3.2.0, metadata-2.0.4
collected 4 items

test_instance06.py::test_method1[10] PASSED
test_instance06.py::test_method1[2] PASSED
test_instance06.py::test_method1[3] PASSED
test_instance06.py::test_method1[4] PASSED

- generated html file: file:///Users/tim/Documents/pyse2023/chapter10/report1.ht
ml -
============================ 4 passed in 0.02s ============================
```

图 10.14

用浏览器打开生成的测试报告文件,截图如图 10.15 所示。

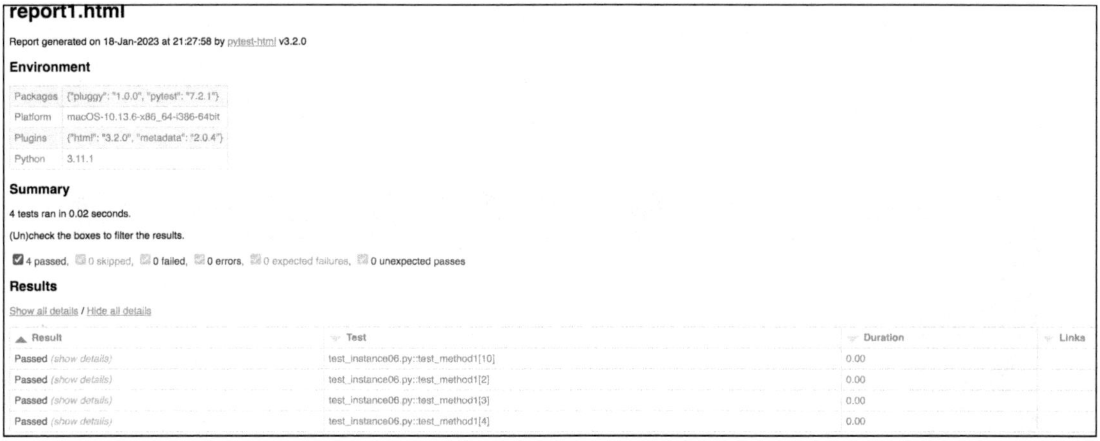

图 10.15

10.1.10 集成 Allure 报告

Allure 可以给用户提供比较直观的测试图文报表，效果比较炫酷。首先需要安装 Allure 软件包，安装命令为"pip install allure-pytest"。

使用 Allure 来执行并获取 test_instance06.py 文件的测试报告，和 pytest-html 进行对比。

首先需要执行测试，获取 JSON 格式的测试结果。

执行测试命令"pytest --alluredir=./reports/allure_report -sv test_instance06.py"，其中的 alluredir 用于设置保存测试结果的路径。测试完成后，JSON 格式的测试结果如图 10.16 所示。

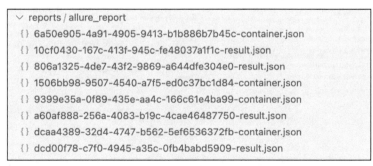

图 10.16

启动 Allure 服务，在启动 Allure 之前，需要先在其官网下载 Allure 软件，下载完成后，执行命令"<path of allure path>/allure serve reports/allure_report"以启动服务，然后打开浏览器，生成测试报告的地址为"http://192.168.0.102:*****/index.html"，报告如图 10.17 所示。

图 10.17

单击 Suites 模块中的 Show all 按钮，会展示一些测试细节，如图 10.18 所示，会自动跳转到 Suites 页面。

图 10.18

10.2 使用 pytest 重构项目

以上学习了 pytest 框架的基本用法，接下来用 pytest 框架改写测试框架，仅需要修改 Base 类与 TestCases。在 Base 目录中新增以下文件，与 UnitBase 类似，执行初始化驱动与驱动退出操作：

```python
import pytest
from selenium import webdriver
from selenium.webdriver.chrome.service import Service
from common.util import projectpath,configAdress
class PyBase:
#driver 存放在项目路径下
    def setup_class(self):
        path = Service(projectpath() + "dr\\chromedriver.exe")
        self.driver = webdriver.Chrome(service=path)
        self.driver.get(configAdress("testUrl", "url"))
    def teardown_class(self):
        self.driver.quit()
        print("test over")
```

在 TestCases 目录中新增测试脚本，运行 pysuite.py 文件，具体代码如下：

```python
from os.path import dirname
import sys,pytest
path = dirname(__file__)
sys.path.append(path.split("testcases")[0])
from Common.function import projectpath
path = projectpath()+"TestCases\\"
report =projectpath()+"Reports\\"
pytest.main(["-s", "--alluredir={}".format(report),path])
#运行 cases 下的所有测试用例
if __name__ == '__main__':
    print(path)
    print(report)
```

测试书写格式与 UnitTest 类似，仅仅是断言格式不同，代码如下：

```
from base.pybase import PyBase
from pageobject.LoginPage import LoginPage
from pageobject.RolePage import NewRole
from pageobject.SysPage import SysPage
from common.ExcelUtil import ExcelUtil
from common.util import projectpath
import time
path = projectpath()+"data\\"+"testdata.xlsx"
ex = ExcelUtil(path,"role")
dic = ex.ExcelDic()
class Testlogin(PyBase):
    def test_01(self):
        login = LoginPage(self.driver)
        res = login.login_system(self.data.get(1)[0],self.data.get(1)[1])
        self.assertIn("",res)
    def test_02(self):
        index = IndexPage(self.driver)
        res = index.position_menu()
        self.assertIn("", res)
    def test_03(self):
        add = AddPositionPage(self.driver)
        res = add.add_post(self.data.get(1)[0],self.data.get(1)[1])
        self.assertIn("", "")
```

本章用 pytest 与 Allure 重构了测试框架，完整代码参见配套资源包。

第 11 章
行为驱动测试

行为驱动开发（Behavior-Driven Development，简称 BDD）的概念在国内测试领域还不怎么流行，应用面也不是特别广。行为驱动是运用结构化的自然语言描述测试场景，并将这些结构化的自然语言转化为可执行的测试脚本或者其他形式。BDD 的优势是，建立了一种"通用语言"，而这种通用语言可以同时被客户和开发者使用。对自动化测试人员来说，掌握了 BDD 之后，可提升测试团队的自动化测试程度，因为 BDD 不会关注程序中相关对象的细节，而只会关注功能点，所以可以减轻回归测试任务的压力。从本章开始，笔者将介绍行为驱动，从环境准备开始到实例演示，逐步深入。

11.1 安装环境

"兵马未动，粮草先行。"首先需要把 BDD 环境准备好，需要安装模块 behave，安装步骤与在 Python 环境中安装其他模块的方式一样，执行命令"pip install behave"。模块安装好之后就可以通过执行命令"pip list"来查看 behave 模块是否出现在已安装列表中。

11.2 行为驱动之小试牛刀

小试牛刀的目的是，根据一个小案例来了解行为驱动的运行模式及简单的应用规则。本节案例是 behave 官方给出的一个例子。笔者画了一个项目目录结构图，方便读者对整个案例进行了解及把握，如图 11.1 所示。

图 11.1

在开始了解具体的 BDD 案例代码之前，可以先熟悉一下 BDD 的一些关键字，具体如下。

（1）Given，表示"假设"，即设置一些前置条件，如在 BDD 之前，假设 behave 模块已经安装等。也可以将其理解成在用户或外部系统等对应用进行交互（操作）前，需要将系统置于一个已知状态（如系统已安装 behave 模块等）。

（2）When，表示"当"，从字面意思上理解其是对条件进行判断的意思，也就是此时或者当某种条件被满足时，用户或外部系统所采取的与被测系统的交互步骤。交互步骤能改变系统的状态（与系统真实地产生了交互）。

（3）And，表示"和"，与 When 关键字搭配使用。

（4）Then，表示"那么"，也就是待观察的结果或者期望的结果。

通过以上对 behave 模块 Scenario（场景）中关键字的描述，会让读者产生 BDD 与自然语言很像的感觉。场景描述文字如图 11.2 所示，文件名为"example.feature"。下面介绍 behave 使用步骤。

第一步，场景描述功能点（Feature），表示脚本用于展示 behave 模块的用法。场景目的是"Run a simple test"，假设为"we have behave installed"，如果"we implement 5 tests"，那么"behave will test them for us!"。通过以上对场景使用的关键字的描述，以及对具体场景的分析，相信读者能够熟悉这种描述测试场景的方式。

```
example.feature
1    Feature: Showing off behave
2
3        Scenario: Run a simple test
4            Given we have behave installed
5            When we implement 5 tests
6            Then behave will test them for us!
```

图 11.2

第二步，为以上类似自然语言一样的场景描述编写代码，将其转换成可以运行的基于行为驱动的测试脚本。文件名为"example_steps.py"，存放在 steps 包下。代码中的函数通过 assert 语句进行断言，需要注意的是，参数 context 是全局变量，它可以被程序中的所有对象或者函数调用，在行为驱动中有承上启下的作用。另外，需要导入 behave 模块中的 Given、When、Then、Step 等功能。代码如下：

```
#coding=utf-8
from behave import given, when, then, step
@given('we have behave installed')
def step_impl(context):
    pass
#数字类型 number 将被转换成整数类型
#以下函数的功能是获取场景文件中设置的数字 5，然后做出判断等
@when('we implement {number:d} tests')
def step_impl(context, number):
    assert number > 1 or number == 0
    context.tests_count = number
@then('behave will test them for us!')
def step_impl(context):
    assert context.failed is False
    assert context.tests_count >= 0
```

第三步，执行脚本。在命令行模式下切换到项目主目录，执行命令"behave"即可，

执行结果如图 11.3 所示。从结果可以发现，1 个功能通过，1 个场景通过，3 个步骤测试通过。

```
Feature: Showing off behave # example.feature:1

  Scenario: Run a simple test          # example.feature:3
    Given we have behave installed     # steps/example_steps.py:3 0.000s
    When we implement 5 tests          # steps/example_steps.py:8 0.000s
    Then behave will test them for us! # steps/example_steps.py:13 0.000s

1 feature passed, 0 failed, 0 skipped
1 scenario passed, 0 failed, 0 skipped
3 steps passed, 0 failed, 0 skipped, 0 undefined
Took 0m0.000s
```

图 11.3

第四步，可以将以上步骤理解为一个正向测试用例。下面以一个反向测试用例来验证 behave 模块的测试输出结果。修改 step 方法中的断言条件，在代码文件 example_steps.py 中将断言部分的 assert number > 1 更改为"assert number > 10"，更改之后的代码文件内容如下，执行结果如图 11.4 所示。

```
#coding=utf-8
from behave import given, when, then, step
@given('we have behave installed')
def step_impl(context):
    pass
#注意，以下的实现函数在校验语句中将>1 更改为>10
@when('we implement {number:d} tests')
def step_impl(context, number):
    assert number > 10 or number == 0
    context.tests_count = number
@then('behave will test them for us!')
def step_impl(context):
    assert context.failed is False
    assert context.tests_count >= 0
```

```
Feature: Showing off behave # example.feature:1

  Scenario: Run a simple test          # example.feature:3
    Given we have behave installed     # steps/example_steps.py:3 0.000s
    When we implement 5 tests          # steps/example_steps.py:8 0.000s
      Traceback (most recent call last):
        File "/Users/jason118/.virtualenvs/selenium4.0-automation/lib/python3.7/site-packages/behave/model.py", line 1329, in run
          match.run(runner.context)
        File "/Users/jason118/.virtualenvs/selenium4.0-automation/lib/python3.7/site-packages/behave/matchers.py", line 98, in run
          self.func(context, *args, **kwargs)
        File "steps/example_steps.py", line 11, in step_impl
          assert number > 10 or number == 0
      AssertionError

    Then behave will test them for us! # None


Failing scenarios:
  example.feature:3  Run a simple test

0 features passed, 1 failed, 0 skipped
0 scenarios passed, 1 failed, 0 skipped
1 step passed, 1 failed, 1 skipped, 0 undefined
Took 0m0.001s
```

图 11.4

11.3 基于 Selenium 的行为驱动测试

通过以上案例中对正反测试用例的讲解，大家对行为驱动测试有了初步的认识。将这种测试机制与 Selenium 框架融合并运用到真正的自动化项目中，将体现更大的价值。笔者将继续以示例来演示这种设想。下面以大牛测试系统登录场景为例来进行讲解。

项目目录结构与上例一致，此处不再赘述。

第一步，场景描述。场景功能是实现登录。可以将功能拆分为 3 步操作：打开登录页面、输入用户名、输入密码。具体可以参考场景文件 example.feature，如图 11.5 所示。

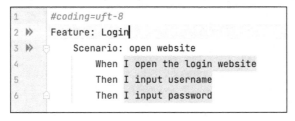

图 11.5

第二步，按照以上对场景的描述，创建如下行为驱动脚本。在脚本中实现了 3 步操作，分别是打开登录页面、输入用户名、输入密码：

```python
#coding=utf-8
from behave import *
from selenium import webdriver
from selenium.webdriver.chrome.service import Service
from selenium.webdriver.common.by import By
#以下函数用于实现打开网站的操作
@when('I open the login website')
def step_impl(context):
    #请在下列代码中添加真实的chromedriver的路径
    chrome_driver_server = Service("/Chapter11.3/s/chromedriver")
    context.driver  = webdriver.Chrome(service=chrome_driver_server)
    context.driver.get('http://localhost/login')
#输入用户名
@Then('I input username')
def step_i2(context):
    context.driver.find_element(By.NAME,"username").send_keys("admin")
#输入密码
@Then('I input password')
def step_i3(context):
    context.driver.find_element(By.NAME,"password").send_keys("admin123")
```

第三步，在项目主目录下执行命令 "behave"，查看测试结果，如图 11.6 所示。

```
Feature: Login # example.feature:2

  Scenario: open website            # example.feature:3
    When I open the login website   # steps/example_steps.py:7 2.102s
    Then I input username           # steps/example_steps.py:15 0.161s
    Then I input password           # steps/example_steps.py:21 0.123s

1 feature passed, 0 failed, 0 skipped
1 scenario passed, 0 failed, 0 skipped
3 steps passed, 0 failed, 0 skipped, 0 undefined
Took 0m2.386s
```

图 11.6

11.4 结合 Page Object 的行为驱动测试

使用 PO 的思想重构或管理行为驱动测试能使测试更有效率。下面以大牛测试系统登录场景为例，整体项目目录结构如图 11.7 所示。

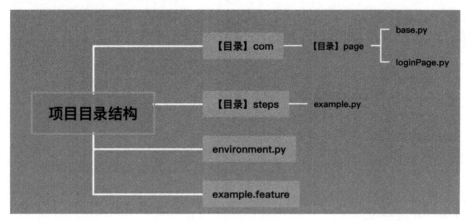

图 11.7

（1）在项目中体现 PO 思想的是 page 目录下的两个文件，分别为 base.py 和 loginPage.py 文件。其中 base.py 文件代码抽取了一些基本方法，如定位元素的方法、打开网站的方法及得到当前网页标题的方法等。代码如下：

```
#coding=utf-8
from selenium.webdriver.common.by import By
class Base:
    def __init__(self,driver):
        self.driver = driver
     #*loc 指传入的是不定参数，意思是 findele 方法可以传入 1 个参数，也可以传入 2 个参数，等等
    def findele(self,*loc):
        return self.driver.find_element(*loc)
    def get_url(self,url):
        self.driver.get(url)
    def get_title(self):
        return self.driver.title
```

（2）page 目录下的另一个文件 loginPage.py 的功能是封装登录页面的操作，这些操作以类的成员方法展现，代码如下：

```python
#coding=utf-8
from features.com.page.base import Base
from selenium.webdriver.common.by import By
#loginPage 继承自 Base 类
class loginPage(Base):
#以下为类的初始化方法，其又调用了父类的初始化方法，这样做的目的是
#将 context.driver 串起来，在调用 PO 类时可以使用超级全局变量 context 下的 driver 对象
    def __init__(self,context):
        super(loginPage,self).__init__(context.driver)
    def login(self,username,password):
        self.findele(By.NAME,"username").clear()
        self.findele(By.NAME,"username").send_keys(username)
        self.findele(By.NAME,"password").clear()
        self.findele(By.NAME,"password").send_keys(password)
```

（3）项目主目录下有 environment.py 文件，该文件的功能是配置行为驱动环境，以便被全局调用，代码如下：

```python
#coding=utf-8
from selenium import webdriver
from selenium.webdriver.chrome.service import Service
def before_all(context):
    chrome_driver = Service("/selenium4.0-automation/ Chapter11.4/chromedriver")
    context.driver = webdriver.Chrome(service=chrome_driver)
def after_tag(contex):
    context.driver.quit()
```

（4）项目主目录下有 example.feature 文件，该文件定义了行为驱动要执行的场景描述细节，代码如下：

```
Feature: Login
Scenario:open website
    When I open the login website "https://localhost/login?"
    Then I input username "admin" and password "admin123"
```

（5）在目录 steps 中有 example.py 文件，该文件定义了行为驱动场景的实现过程，代码中涉及正则表达式的使用，具体代码如下：

```
#coding=utf-8
from behave import *
from features.com.page.loginPage import loginPage
# "re" 表示在 steps 中定义正则表达式
use_step_matcher('re')
#抓取场景文件 example.feature 中的 URL 值，将其传给 URL，然后执行下面的操作
@when('I open the login website "([^"]*)"')
def step_reg(context,url):
    loginPage(context).get_url(url)
@Then('I input username "([^"]*)" and password "([^"]*)"')
def step_r(context,username,password):
    loginPage(context).login(username,password)
```

（6）执行测试，测试结果如图 11.8 所示。

```
Feature: Login  # example.feature:1

  Scenario: open website                                    # example.feature:2
    When I open the login website                           # steps/example.py:10 0.402s
    Then I input username "admin" and password "admin123"   # steps/example.py:14 0.277s

1 feature passed, 0 failed, 0 skipped
1 scenario passed, 0 failed, 0 skipped
2 steps passed, 0 failed, 0 skipped, 0 undefined
Took 0m0.679s
```

图 11.8

以上对行为驱动模式大牛测试系统登录进行了流程和结构的梳理，通过对总体结构的讲解和对细节的解释，相信大家对行为驱动测试会产生更深刻的认识。

第四篇

平 台 篇

通过前面的学习，基本框架已经形成，但是这在实际项目中还不够，因为实际项目可能很复杂，需要频繁地进行迭代回归测试，这时就需要建立平台进行管理了。如持续集成工具 Jenkins 可以将测试流程自动化（如自动部署 App、自动规划测试执行步骤等）；如针对邮件服务器可以设置自动发送邮件功能（测试执行完成或者测试执行异常等）；如代码托管工具 Git（可以对测试脚本、测试数据进行版本控制，方便管理）及 Docker 容器技术。搭建平台的优点如下。

- 易于管理账号。
- 可以集成不同的环境。
- 存储每次的执行结果（数据库存储等模式）。
- 发送邮件。
- 实现定时执行测试。

本篇对应的章节如下。

第 12 章　测试平台维护与项目部署

第 13 章　Docker 容器技术与多线程测试

第 12 章
测试平台维护与项目部署

本章将介绍与平台相关的一些知识,涉及 Git 和 Jenkins 应用,它们都是当前搭建测试平台时经常使用的技术。

12.1　Git 应用

Git 是分布式版本控制系统,是当前比较流行、好用的一款版本控制系统。使用 Git 可以有效、高速地处理项目版本管理工作。Git 是 Linus Torvalds 为了管理 Linux 内核开发而开发的版本控制软件。

下面举个例子来说明版本控制的重要性。当需要自动化部署 Tomcat 应用,设置 Tomcat 的配置时,如果没有进行版本控制,我们就可能采取备份/复制等传统方式来保持对配置的追踪和更新,这种方式会显得版本特别混乱且维护也不方便,增加了维护成本。而如果使用 Git,可以减少这种重复劳动,更方便追踪版本情况,也可以随时切换版本。

以下为利用 Git 进行开发的经典场景。

(1) 先从 Git 服务器上克隆完整的 Git 仓库(包括代码和版本信息)到本地。

(2) 在本地的 Git 环境下根据不同的开发目的,创建分支,修改代码。

(3) 在本地提交自己的代码给自己创建的分支。

(4) 在本地合并分支。

(5) 把服务器上最新版的代码从远程分支拉取(fetch)下来,然后与本地主分支合并。

(6) 生成补丁(Patch),把补丁发送给主开发者。

(7) 根据主开发者的反馈,如果主开发者发现两个开发者之间有冲突(可以通过合作解决的冲突),就会要求他们先解决冲突,再由其中一个人提交版本。如果主开发者可以自己解决冲突或者没有冲突,则通过。

(8) 解决冲突,开发者之间可以使用 pull 命令来解决冲突(pull 命令的功能是先从 Git 仓库中抓取最新的代码到本地),解决完冲突之后再向主开发者提交补丁。

12.1.1 安装 Git

1. Linux Git

针对不同的 Linux 版本的操作系统,可以使用包管理工具进行 Git 的快捷安装。如 Debian/Ubuntu,可以使用命令 "apt-get install git" 直接安装;最新版的 Fedora,可以使用命令 "dnf install git" 直接安装,具体如图 12.1 所示。

2. Windows Git

Windows 操作系统上的 Git 安装更容易一些。如图 12.2 所示,下载 32 位或 64 位基于 Windows 图形化界面的安装包,具体安装过程如下。

在安装过程中,可以按照提示进行安装。需要设置 Git 的默认编辑器(如图 12.3 所示),选择本机已经安装的编辑器即可,若本地安装了 Notepad++,则选择 Notepad++。

第 12 章　测试平台维护与项目部署

```
Download for Linux and Unix

It is easiest to install Git on Linux using the preferred package manager of your Linux distribution. If you
prefer to build from source, you can find the tarballs on kernel.org.

Debian/Ubuntu
For the latest stable version for your release of Debian/Ubuntu
    # apt-get install git
For Ubuntu, this PPA provides the latest stable upstream Git version
    # add-apt-repository ppa:git-core/ppa  # apt update; apt install git
Fedora
    # yum install git (up to Fedora 21)
    # dnf install git (Fedora 22 and later)
Gentoo
    # emerge --ask --verbose dev-vcs/git
Arch Linux
    # pacman -S git
openSUSE
    # zypper install git
Mageia
    # urpmi git
Nix/NixOS
    # nix-env -i git
FreeBSD
    # pkg install git
Solaris 9/10/11 (OpenCSW)
    # pkgutil -i git
Solaris 11 Express
    # pkg install developer/versioning/git
OpenBSD
```

图 12.1

```
Download for Windows

Click here to download the latest (2.39.2) 64-bit version of Git for Windows. This is the most
recent maintained build. It was released 4 days ago, on 2023-02-14.

Other Git for Windows downloads
Standalone Installer
32-bit Git for Windows Setup.

64-bit Git for Windows Setup.

Portable ("thumbdrive edition")
32-bit Git for Windows Portable.

64-bit Git for Windows Portable.
```

图 12.2

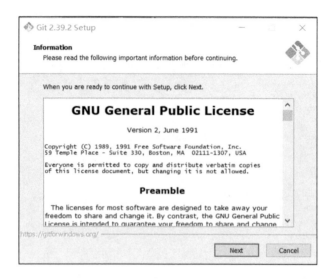

图 12.3

安装完成后，在 Windows 桌面上单击鼠标右键，将出现如图 12.4 所示的菜单，选择"Git GUI Here"，打开 Git Gui 界面，如图 12.5 所示。也可以选择"Git Bash Here"，用法与在 Linux 系统下操作 Git 类似，也是命令行方式的。

图 12.4

图 12.5

3. macOS Git

macOS Git 安装包如图 12.6 所示，选择 macOS X 版本的进行下载，再下载 dmg 安装包进行安装。macOS 自带 Git，一般不需要安装，除非版本不合适，这时可到其官网下载最新版，按照提示一步一步安装即可。

图 12.6

安装成功后，在 macOS 中运行命令"git--version"，可以显示 Git 的版本信息，就说明 Git 安装成功。

12.1.2　Git 常用操作

Git 常用操作介绍如下。

（1）Git 仓库一般是用 GitHub 应用程序管理的。用 Git 管理项目，需要将 Git 仓库中相应的分支克隆到本地，比如执行命令"git clone -b master https://******.com/user/ProjectName.git"。git clone 命令的功能是克隆 Git 仓库中的内容到本地当前目录，-b master 参数指当前克隆的是 master 分支的内容。如图 12.7 所示，master 分支已被成功克隆到本地。

```
sh-3.2# git clone -b master https://        /ld0906/AutoTest.git
Cloning into 'AutoTest'...
remote: Enumerating objects: 24, done.
remote: Counting objects: 100% (3/3), done.
remote: Compressing objects: 100% (2/2), done.
remote: Total 24 (delta 0), reused 3 (delta 0), pack-reused 21
Unpacking objects: 100% (24/24), 4.22 KiB | 288.00 KiB/s, done.
sh-3.2#
```

图 12.7

（2）将本地新增的文件/文件夹上传到 Git 仓库（远端）。本地新增了两个文件，要如何正确地将其上传到 Git 远端库呢？以下为具体步骤。

① 更新本地库，命令为"git pull"，目的是在提交本地变更前将本地库更新到最新的状态。

② 运行命令"git status"，查看当前库的变更，会列出所有变更，包括新增、修改、删除等变更的文件。如图 12.8 所示，"newfile1.log"和"newfile2.log"是本地新增的文件。

```
sh-3.2# git status
On branch master
Your branch is ahead of 'origin/master' by 1 commit.
  (use "git push" to publish your local commits)

Untracked files:
  (use "git add <file>..." to include in what will be committed)
        newfile1.log
        newfile2.log

nothing added to commit but untracked files present (use "git add" to track)
```

图 12.8

③ 上传文件 newfile1.log 需要运行命令"git add newfile1.log"；上传当前所有的变更，则运行命令"git add -A"；如果只添加新文件或更改过的文件，但不包括删除的文件，则运行命令"git add ."；如果只添加编辑过的文件或删除的文件，但不包括新添加的文件，则运行命令"git add -u"。运行"git add"命令之后，就将代码从工作区添加到了暂存区。

④ 运行命令"git commit"，可以将代码从暂存区提交到本地库。需要添加注解，如图 12.9 所示，说明此次提交的目的和用途等。

```
# Please enter the commit message for your changes. Lines starting
# with '#' will be ignored, and an empty message aborts the commit.
#
# On branch master
# Your branch is ahead of 'origin/master' by 1 commit.
#   (use "git push" to publish your local commits)
#
# Changes to be committed:
#       new file:   newfile1.log
#       new file:   newfile2.log
#
~
```

图 12.9

⑤ 运行命令"git push"，目的是推送本地库中的当前分支到远程服务器上的对应分支。执行命令后新增的文件从本地提交到了远程服务器上。

（3）假设在 Git 本地工作区修改了一个文件，在将其提交到远程服务器之前，想查看该文件的变更情况，这时可以运行命令"git diff newfile1.log"。运行命令的结果如图 12.10 所示，控制台上添加了一行字符"newadd23"。

```
[sh-3.2# git diff newfile1.log
diff --git a/newfile1.log b/newfile1.log
index ccfe4ce..d9e0146 100644
--- a/newfile1.log
+++ b/newfile1.log
@@ -1 +1,2 @@
 newadd23
+newadd23
```

图 12.10

（4）Git 回退版本的操作。回退版本的操作在工作中也比较常见。回退到上一个版本，可运行命令"git reset --hard HEAD^"；回退两个版本，可运行命令"git reset --hard HEAD^^"；那么回退到指定版本该怎么操作呢？可以采取如下步骤。

① 运行 Git 命令"git log"，运行结果如图 12.11 所示（图片中只截取了部分版本信息）。

```
[sh-3.2# git log
commit 36ffdb2450e9439d3abfc863bb6cb08a8c085eb8 (HEAD -> master, origin/master, origin/HEAD)
Author: Id0906 <706808121@qq.com>
Date:   Sun Feb 19 00:37:41 2023 +0800

    newfile1.log

commit eee21b0f1f8202b84af9cad27ce75e59ec6228f8
Author: Id0906 <706808121@qq.com>
Date:   Sat Jan 28 17:34:10 2023 +0800

    newly added two files

commit 1831c12156bc2899194b791c942d23c98ea19a43
Author: jason <jason118@jasondeMacBook-Pro.local>
Date:   Thu Jun 13 22:18:42 2019 +0800

    for testing
```

图 12.11

② 选择要回退的指定版本号，如选择版本"eee21b0f1f8202b84af9cad27ce75e59ec6228f8"，则执行 Git 命令"git reset --hard eee21b0f1f8202b84af9cad27ce75e59ec6228f8"。

12.1.3 运用 GitHub

GitHub 是一个面向开源及私有软件项目的托管平台。因只支持 Git 作为唯一的版本库格式进行托管，故名 GitHub。

GitHub 除了 Git 代码仓库托管及基本的 Web 管理界面，还提供了订阅、讨论组、文本渲染、在线文件编辑器、协作图谱（报表）、代码片段分享（Gist）等功能，其中不乏知名的开源项目 Ruby on Rails、jQuery、Python 等。很多企业在内网中也部署了专属的 GitHub 应用，以方便管理内部代码。

下面以公网 GitHub 为例，介绍如何一步一步将自己的代码上传至 GitHub，再克隆到本地。

（1）在 GitHub 网站上注册 GitHub 账号。

（2）新建 GitHub 仓库，如图 12.12 所示。单击"New"按钮，在新建仓库页面填写仓库的相关信息，设置仓库的权限，可以选择"Public"（任何人都可以看到、下载等）或者"Private"（可以配置哪些人有下载、浏览、提交代码等权限），具体如图 12.13 所示。

图 12.12

（3）单击"Create repository"按钮后，在"Quick setup"页面中有 3 种方式可以快速创建 GitHub 仓库，如图 12.14 所示。一般常用的是第一种，将本地代码传入新仓库中，截图中显示已经列出了每一步的命令操作。有两点需要注意，一是要把当前位置切换到代码

第 12 章　测试平台维护与项目部署

目录，二是如果有很多文件要操作，按照之前 Git 的操作命令，可以使用命令"git add -A"。

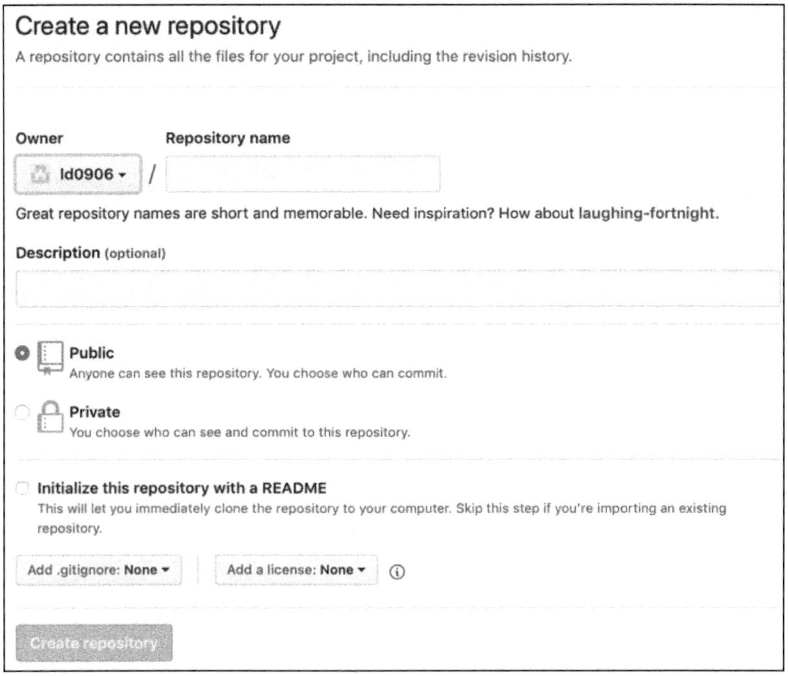

图 12.13

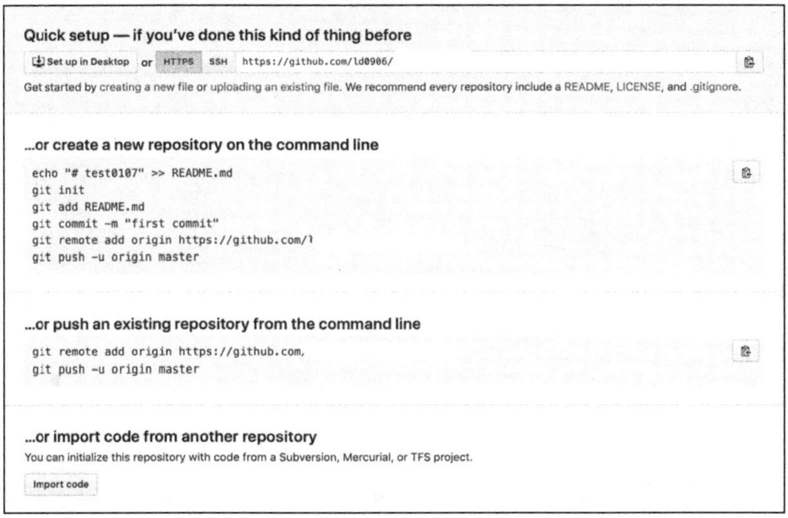

图 12.14

（4）GitHub 仓库创建完成后，如图 12.15 所示。

图 12.15

（5）获取 Git URL，可以通过如图 12.16 所示的方式获取。

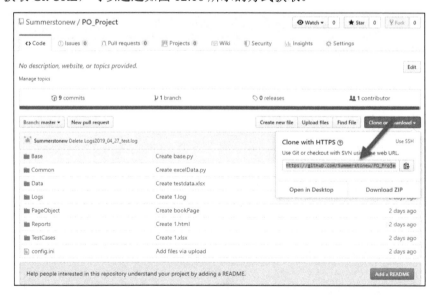

图 12.16

（6）获取 Git URL 之后，使用命令"git clone -b master https://******.com/ld0906/jenkinstest.git"将代码克隆到本地，如图 12.17 所示。

__init__.py	April 5, 2022 at 23:52	Zero bytes	Python Script
__pycache__	Today at 07:56	--	Folder
Base	Today at 07:56	--	Folder
Common	Today at 08:46	--	Folder
config.ini	Today at 07:59	85 bytes	Document
Data	Today at 09:16	--	Folder
Logs	Today at 07:57	--	Folder
PageObject	Today at 09:40	--	Folder
Reports	Today at 09:14	--	Folder
2023-01-28-09_14_04.html	Today at 09:14	8 KB	HTML text
TestCases	Today at 09:17	--	Folder

图 12.17

12.2 安装 Jenkins

Jenkins Job 可以持续地、自动地构建/测试软件项目，可以同时监控一些定时执行的任务。Jenkins 的主要特性如下。

（1）易于安装，将 Jenkins.war 包部署到 Servlet 容器中即可，不需要后端的数据库支持。

（2）易于配置，可以通过自带的 Web 界面方便且直观地设置参数。

（3）可以集成 E-mail 和 JUnit/TestNG 测试报告。

（4）支持分布式构建，可以让多台计算机一起联机部署测试。

（5）支持插件扩展等。

（6）符合 CI/CD（持续集成/持续部署）机制，符合现在主流的开发测试流程。

打开 Jenkins 官网，在其中选择 Windows 版本进行下载，如图 12.18 所示，下载完成后得到的是一个 msi 格式的 Windows 安装包，可以按照安装步骤的提示一步一步地完成安装。

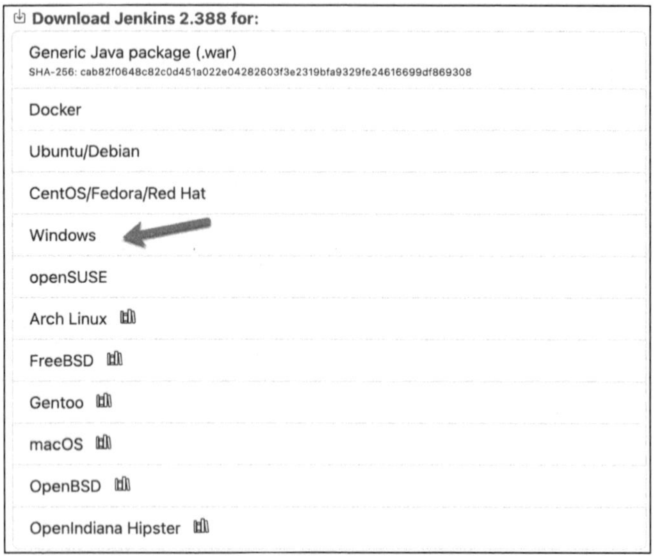

图 12.18

也可以选择 war 版本（Generic Java Package），安装好 JDK 后，在 DOS 窗口中切换到"jenkins.war"目录下，输入命令"java -jar jenkins.war"后便可启动 Jenkins。

war 包安装完成后，可以在浏览器中打开"http://localhost:8081/"，结果如图 12.19 所示，需要解锁 Jenkins，密码在安装过程中已经产生，密码文件的路径是"C:\Users\用户名\.jenkins\secrets\initialAdminPassword"。打开文件"initialAdminPassword"即可获取密码文本，然后完成解锁，并进入下一步。

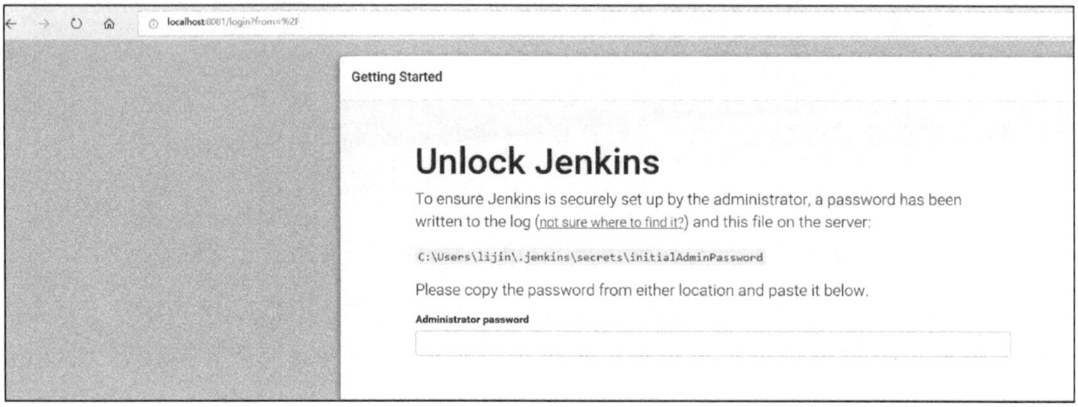

图 12.19

然后是定制化安装 Jenkins 插件，笔者推荐第一种安装模式"Install suggested plugins"，使用这种模式将安装社区推荐的最有价值的插件集合，单击安装模式进行下一步安装，如图 12.20 所示。

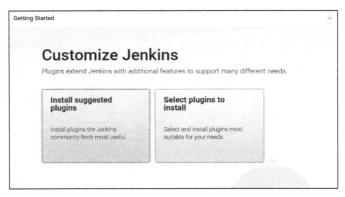

图 12.20

开始安装选择的插件集合，如图 12.21 所示，这个过程要持续一段时间才能完成。

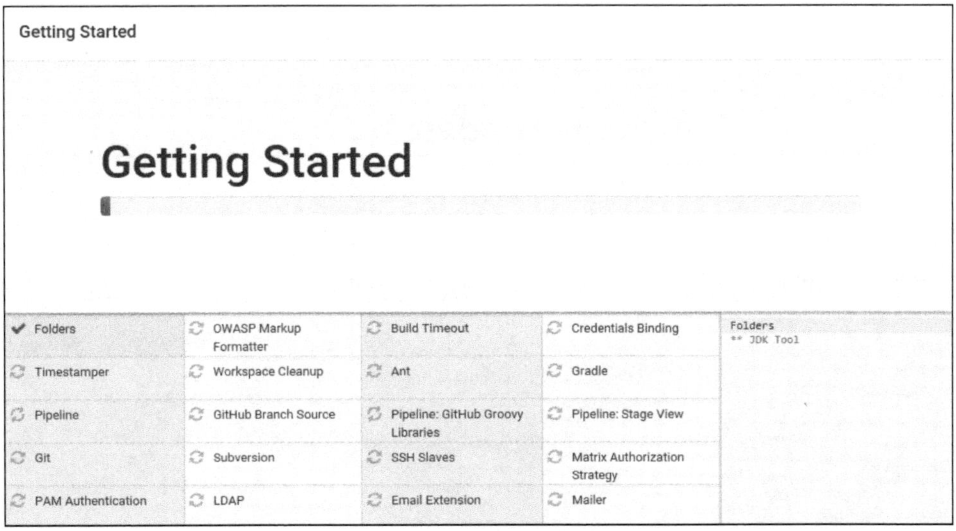

图 12.21

插件集合安装完成后，会自动跳转到下一步，如图 12.22 所示，这时需要设置账户、密码等信息。进入下一步，安装结束，如图 12.23 所示。

安装完毕后，登入 Jenkins 系统，首页如图 12.24 所示，现在就可以开始使用 Jenkins 了。

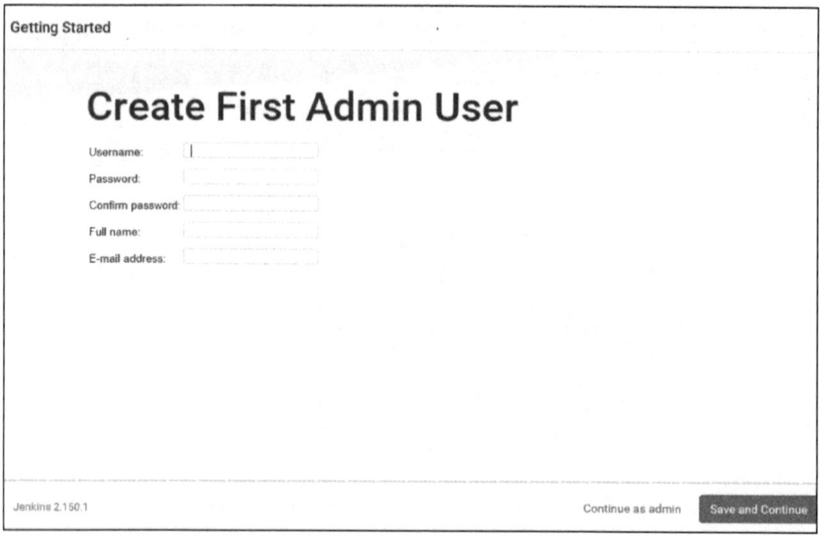

图 12.22

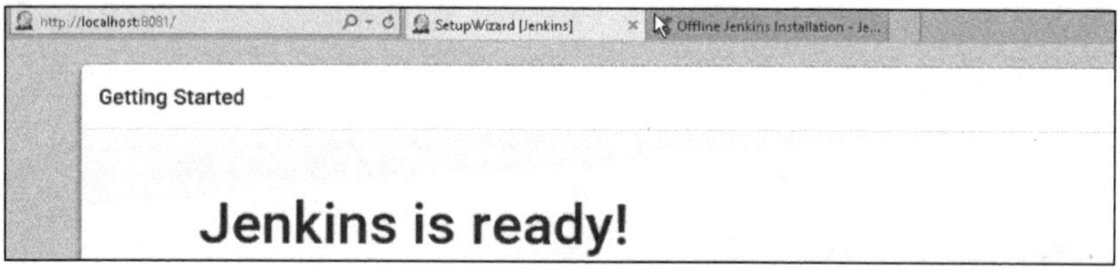

图 12.23

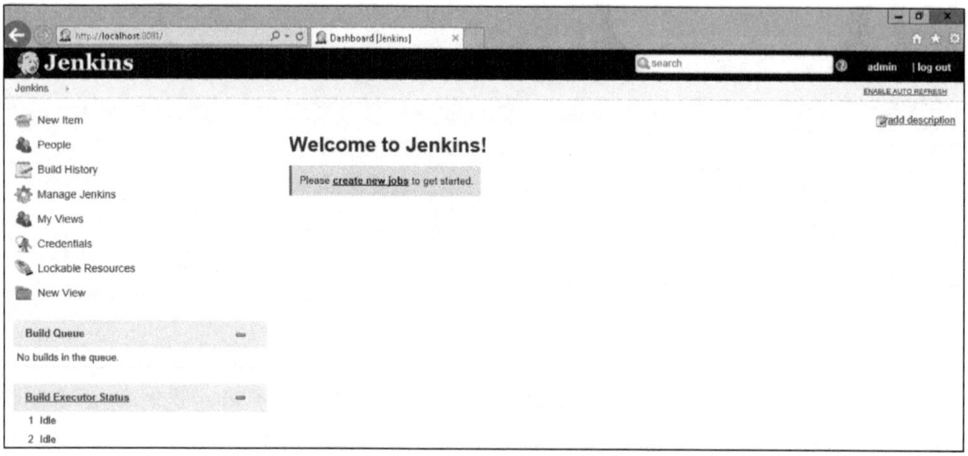

图 12.24

12.3　配置 Jenkins

在配置 Jenkins 之前，需要了解代码编译和构建的一些工具，如 Make、Ant、Maven。

Make 是 Windows 或者 Linux 系统中比较原始的编译工具。Windows 系统下它对应的工具名为"nmake"，主要用于控制编译器的编译过程和连接器的连接过程等。

Make 工具编译一些比较复杂的应用程序时不是很方便，语法较复杂，在这样的情况下就衍生出了 Ant 工具。Ant 工具在软件环境的自动化构建方面使用比较多。

Maven 相当于对 Ant 进行了进一步的优化和改进，它利用 Maven plugin 来完成构建过程，可以控制编译过程和连接过程，可以生成报告并完成一些测试。下面是其中常用的模块功能。

1. Configure Global Security

如图 12.25 所示，在左侧导航栏处选择"Manage Jenkins"，在右侧选择"Configure Global Security"。

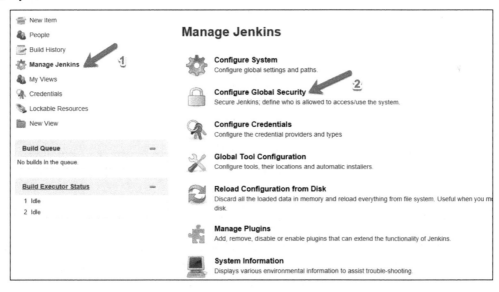

图 12.25

进入 Configure Global Security 界面，勾选"Enable security"选项，授权项也需要设置，如选择"Anyone can do anything"选项，即任何用户都可以做任何事情，表示没有任何限制，如图 12.26 所示。再比如选择"Matrix-based security"选项，表明可以给登录用户或者非登录用户进行权限分配，如图 12.27 所示。

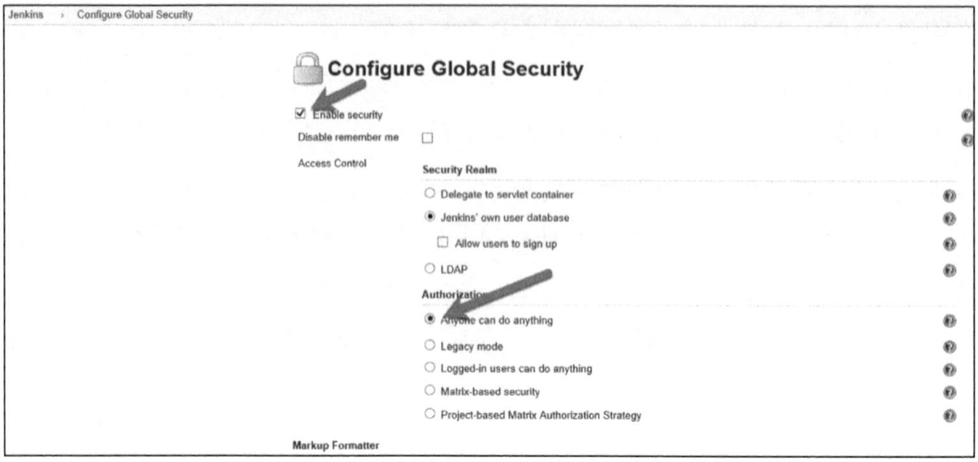

图 12.26

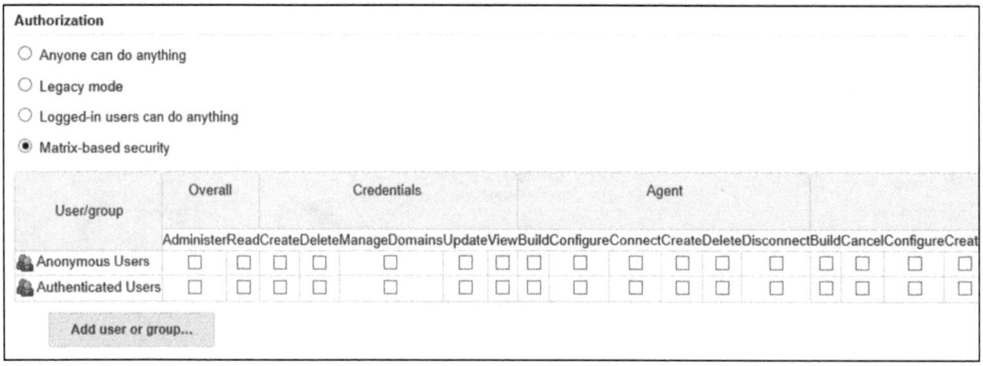

图 12.27

2. 管理插件功能

在 Jenkins 管理界面上可以找到管理插件的功能，如图 12.28 所示，可实现插件安装、插件删除、插件禁用/启用等功能，具体可以参考如图 12.29 所示的 Plugin Manager 界面。

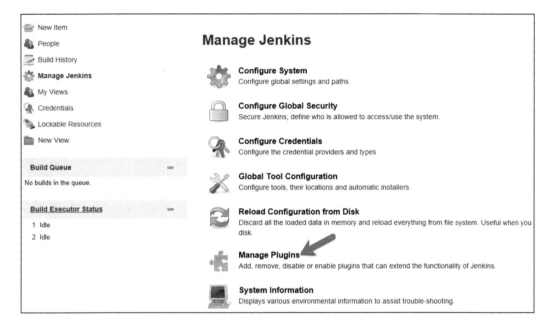

图 12.28

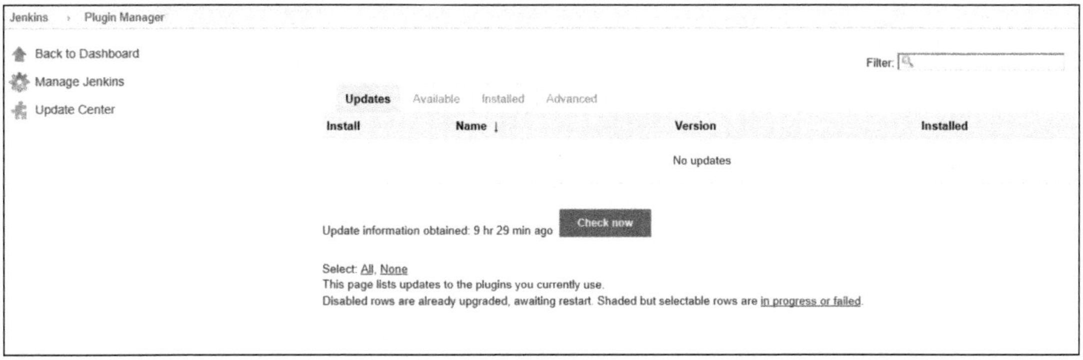

图 12.29

要安装新的插件，可以在 Available 界面查询可用的插件，然后单击"Install without restart"按钮，即可进行安装，如图 12.30 所示。

3. 搭建 Jenkins master-slave

搭建 master-slave 节点相当于串联了多台计算机，可以体现出 Jenkins 的可扩展能力，能最大限度地利用更多的服务器资源，新建节点的方法如图 12.31 所示。

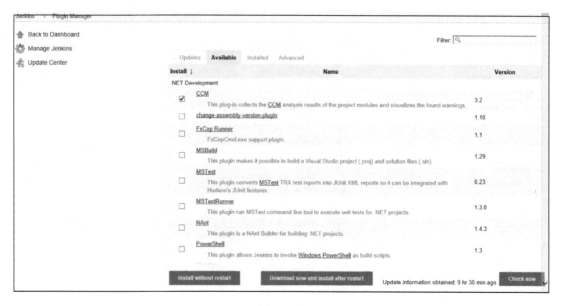

图 12.30

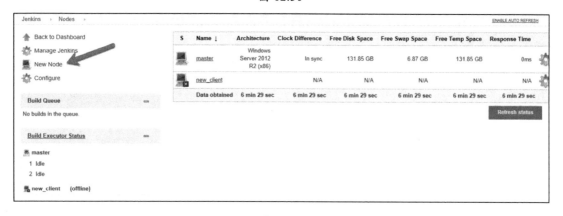

图 12.31

配置节点的步骤是,单击左侧导航栏中的"Configure",在右侧设置"Remote root directory";在"Launch method"中选择"Launch agent via Java Web Start",具体如图 12.32 所示。

以上配置完成后,单击"Save"按钮,跳转到如图 12.33 所示的界面,提示当前的客户端还处于离线状态,需要通过浏览器来将其设置为在线状态。然后单击"Launch"按钮,会弹出如图 12.34 所示的对话框,这时需要运行 Java 应用程序"Jenkins Remoting Agent",单击"Run"按钮。

第 12 章　测试平台维护与项目部署

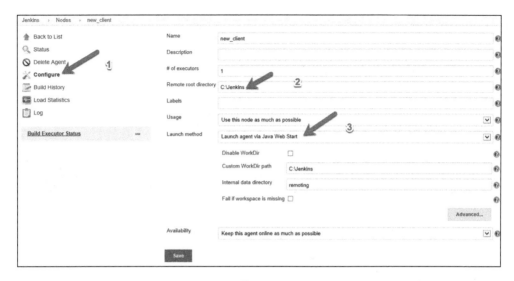

图 12.32

图 12.33

运行成功之后在 Jenkins agent 窗口中会提示"Connected",如图 12.35 所示。

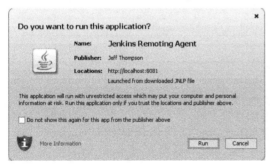
图 12.34

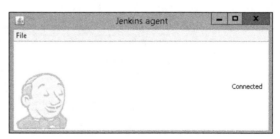
图 12.35

节点创建完毕并启动成功后,浏览如图 12.36 所示的 Nodes 列表,master 节点"master"和 slave 节点"new_client"的状态是在线的。

S	Name ↓	Architecture	Clock Difference	Free Disk Space	Free Swap Space	Free Temp Space	Response Time
🖥	master	Windows Server 2012 R2 (x86)	In sync	131.66 GB	6.73 GB	131.66 GB	0ms
🖥	new_client	Windows Server 2012 R2 (x86)	In sync	131.66 GB	6.73 GB	131.66 GB	116ms
	Data obtained	3 min 7 sec	3 min 7 sec	3 min 7 sec	3 min 6 sec	3 min 7 sec	3 min 7 sec

图 12.36

在安装完 Jenkins 后,其 master 节点是默认配置好的。以上我们已经创建了一个 slave 节点,如果要新建 Jenkins Job,有两种选择,master 节点或 slave 节点。

12.4 Jenkins 应用

12.4.1 自由风格项目介绍

新建 Jenkins Job,如图 12.37 所示,需要输入 Jenkins 的项目名称,我们把它命名为"登录测试",选择"Freestyle project"类别,意思是自由风格项目。

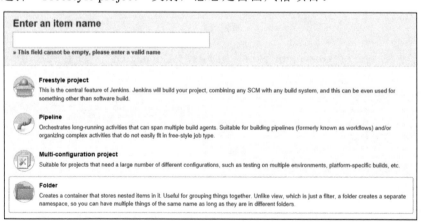

图 12.37

完成以上步骤后，单击界面下方的"OK"按钮，进入 Job 配置界面，如图 12.38 所示。

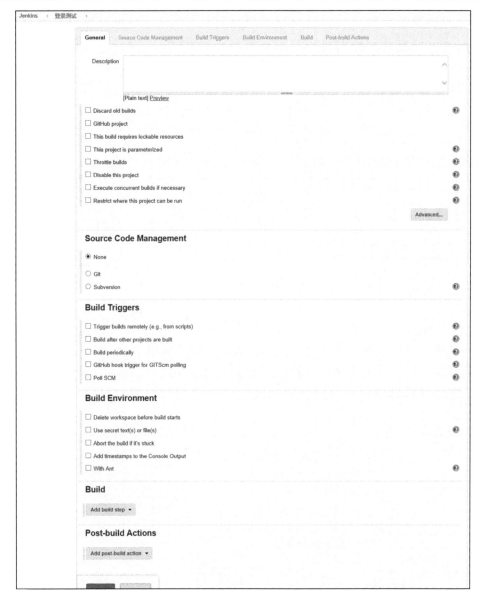

图 12.38

勾选"Restrict where this project can be run"选项，在"Label Expression"下拉列表中选择 Job 运行的节点，在本例中可以选择"master"或者"new_client"，如图 12.39 所示。

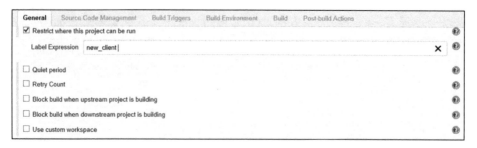

图 12.39

自定义工作空间，如图 12.40 所示，定义的工作目录是"C:\Jenkins_Job"。

图 12.40

代码管理，可以选择从 SVN 或者 Git 上获取代码。如图 12.41 所示，可以设置 Git 仓库地址、验证方式（比如用户名、密码）、选择仓库的分支等。

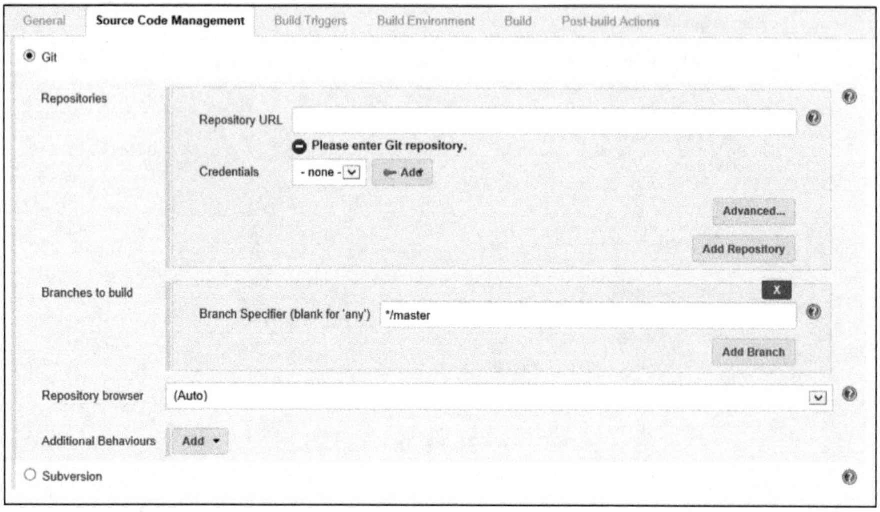

图 12.41

以上步骤设置了环境和代码属性等，接下来需要设置 Job 执行的命令，在"Command"输入框中输入要执行的代码命令，选择"Execute Windows batch command"选项后，如图 12.42 所示，即可使用 Windows 命令执行，至此 Job 构建步骤全部完成。

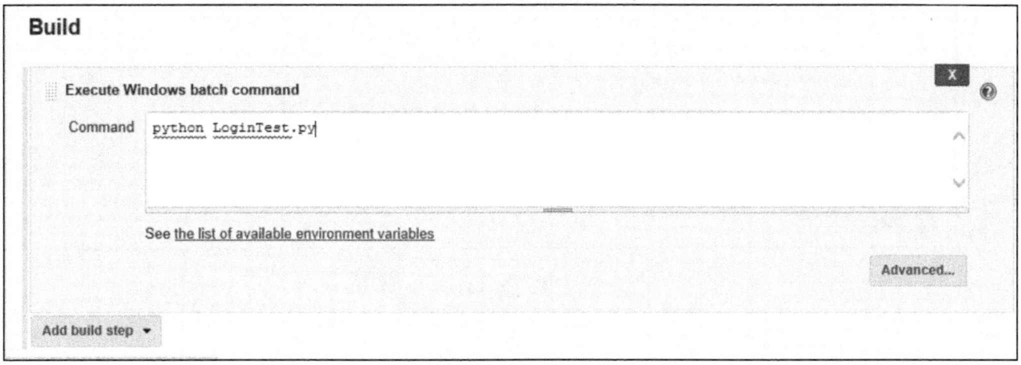

图 12.42

Job 执行完毕后，需要执行一些后续操作。如果要生成 JUnit 的测试结果，可以在"Build"界面中单击"Add post-build action"下拉按钮，然后选择"Publish JUnit test result report"选项进行配置，如图 12.43 所示。如果要发送邮件，则选择"E-mail Notification"选项进行配置，如图 12.44 所示。

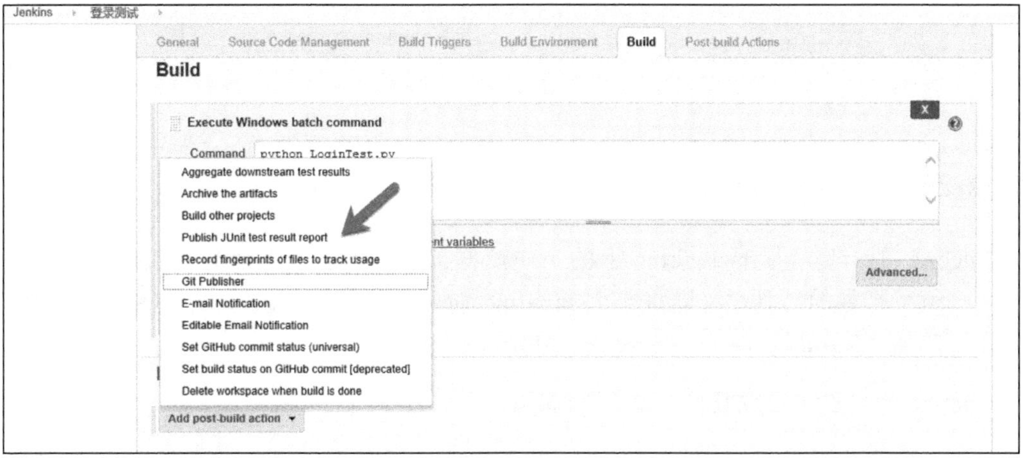

图 12.43

Job 相关的配置都设置好后，运行 Job，如图 12.45 所示，单击左侧导航栏处的"Build Now"即可，此时 Jenkins Job 就可以有效、正确地运转了。

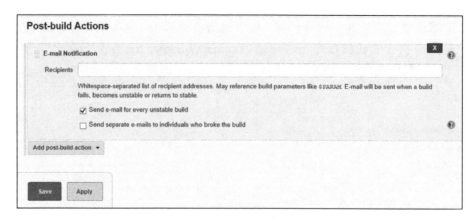

图 12.44

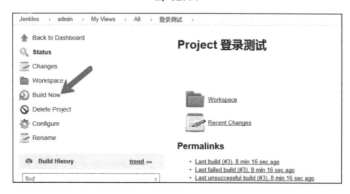

图 12.45

12.4.2　Jenkins Pipeline

Pipeline 是一套运行于 Jenkins 上的工作流框架,可以连接多个节点的任务。Pipeline 提供了一组可扩展的工具,可以将一个 Pipeline 划分成若干个 Stage,每个 Stage 代表了一组操作,比如"Build""Test""Deploy""Post actions"。

Pipeline 和自由风格项目的主要区别如下。

(1)自由风格项目的模式是一种自上而下的 Job 调度,如果有 3 个 Stage,则有 3 个 Job 设定。然后用 Build Flow Plugin 调度多个子 Job。

(2)Pipeline 的模式是在单个 Job 中完成对所有任务的编排,有全局性。比如,可以在一个脚本中实现多个相互关联的步骤,运行结果通常是以通道形式展示的。

Jenkins Pipeline 实践操作的目的是,通过 Pipeline 脚本实现在 GitHub 中抓取测试代码到工作空间中,并开启虚拟 Python 测试环境,最后执行 Selenium 测试。

具体以第 9 章的测试代码为案例进行测试,测试大牛测试系统登录功能,如果成功登录,会在当前目录下生成一个截图"login_screenshot.png"。

(1)在 Jenkins 中创建一个 Pipeline 任务,如图 12.46 所示,在类别中选择"Pipeline"选项,填写任务名"selenium_test",单击界面下方的"OK"按钮。

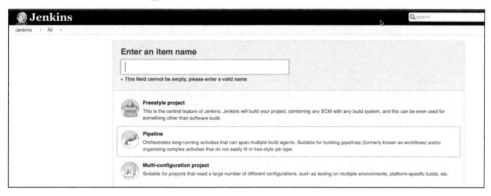

图 12.46

(2)Pipeline 任务的核心就是 Pipeline 脚本设定,如图 12.47 所示,在脚本中需要清晰定义与 Job 相关的 Stage 和每个 Stage 要完成的任务。在"Definition"下拉列表中可以选择"Pipeline script"或者"Pipeline script from SCM",两者的区别是前者脚本存放在本地,后者利用 Git 进行管理。

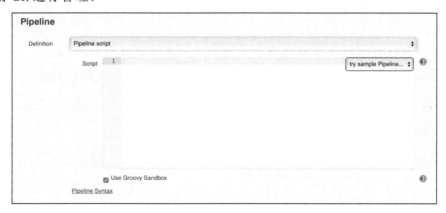

图 12.47

如果大家对 Pipeline 脚本不熟悉，可以参考 Pipeline Syntax，具体可以单击如图 12.48 所示的链接，学习其基础用法。

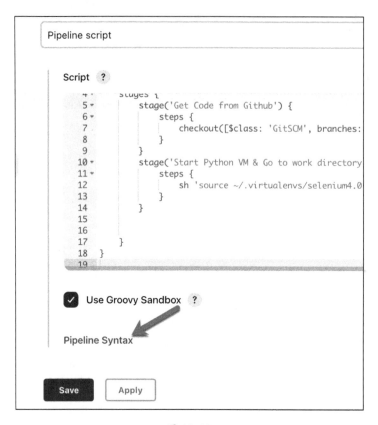

图 12.48

下面开始创作脚本，获取 GitHub 代码时，需要验证，需要设置 Credentials。

第一步，在 Jenkins 管理界面中单击"Manage Jenkins"，在"Security"中选择"Manage Credentials"，然后单击"(global)"，如图 12.49 所示。

在"Global credentials(unrestricted)"界面中，单击"Add Credentials"，具体如图 12.50 所示。

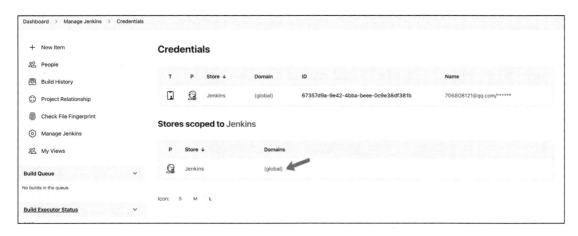

图 12.49

图 12.50

在"New credentials"界面中,在"Kind"下选择"Username with password"。在"Username"和"Password"下输入用户名和密码,最后单击"Create"按钮,具体如图 12.51 所示。至此,Credentials 创建成功。

第二步,获取 GitHub 代码,在"Pipeline Syntax"中的"Snippet Generator"处获取需要的 Pipeline 脚本,在"Sample Step"下选择"checkout: Check out from version control",在"SCM"下选择"Git",输入 Repository URL,具体如图 12.52 所示。

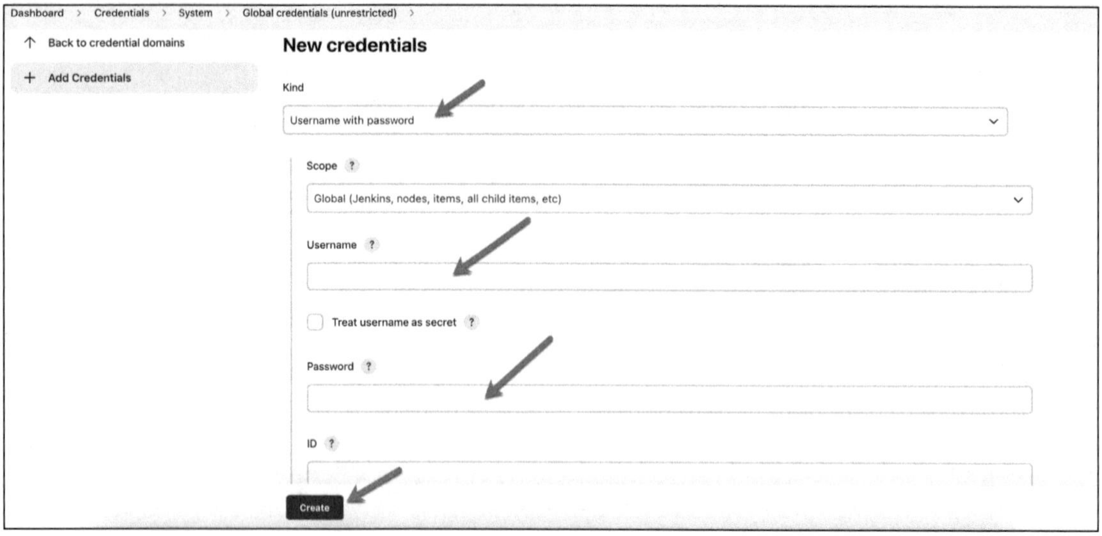

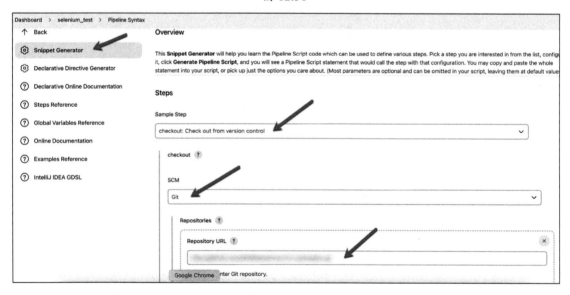

图 12.51

图 12.52

选择之前设置的 Credentials,选择需要抓取的分支,比如"main",具体如图 12.53 所示。

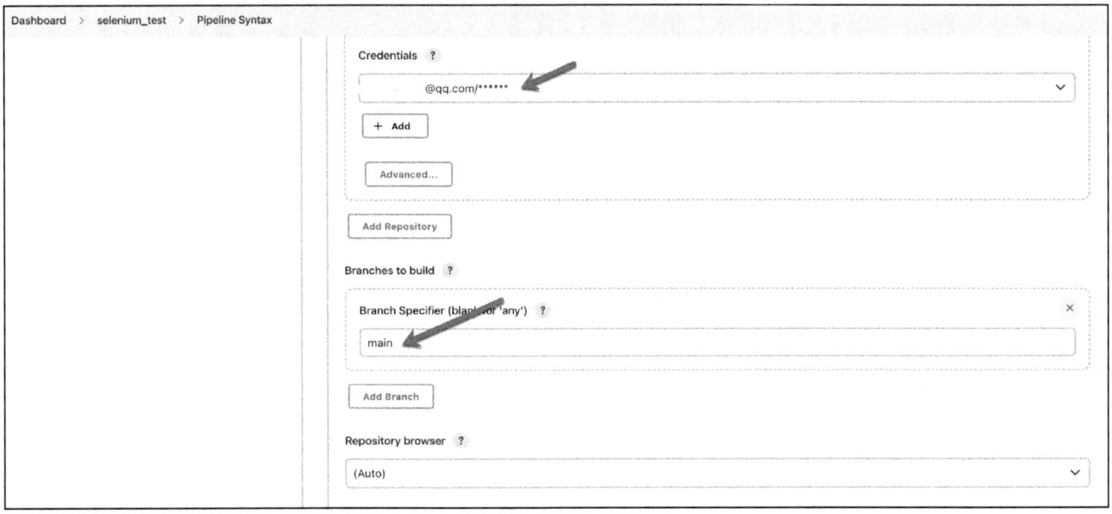

图 12.53

最后单击"Generate Pipeline Script"按钮,至此成功生成了抓取 GitHub 代码的脚本,具体如图 12.54 所示,脚本内容如下:

```
checkout([$class: 'GitSCM', branches: [[name: 'main']], extensions: [],
userRemoteConfigs: [[credentialsId: '67357d9a-9e42-4bba-beee-0c9e38df381b',
url: 'https://******.com/ld0906/selenium4.0-automation.git']]])
```

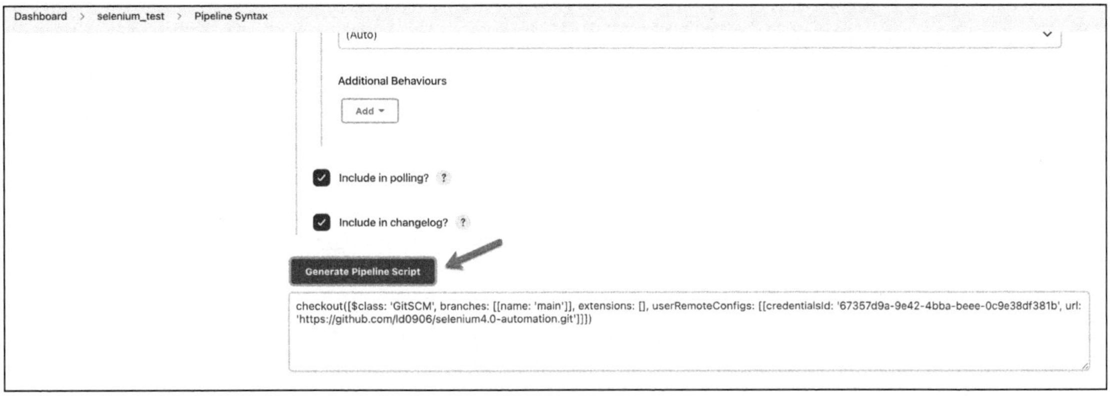

图 12.54

第三步,开启虚拟 Python 测试环境,然后执行 Selenium 测试。此时需要获取执行 Shell 脚本的 Pipeline 脚本。在"Sample Step"下选择"sh: Shell Script",在"Shell Script"输

入框中输入要执行的 Shell 脚本，如图 12.55 所示。Shell 脚本语句之间最好用&&连接，因为在开启虚拟 Python 测试环境后，会在断开虚拟环境后再执行其他 Shell 脚本，Pipeline 脚本如下：

```
sh 'source ~/.virtualenvs/selenium4.0-automation/bin/activate && python -V && cd Chapter9 && python test_system.py'
```

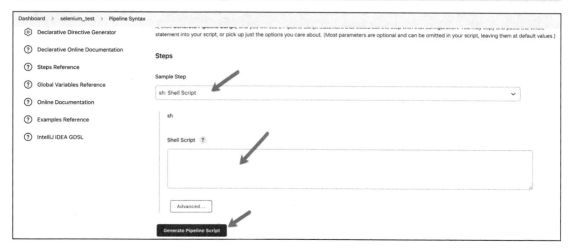

图 12.55

至此 Pipeline 脚本的讲解已经完成。下面对脚本总体布局进行讲解，具体如图 12.56 所示。"agent any"表示在任何可用的节点上执行 Pipeline，"stages"表示 Pipeline 涉及的所有阶段，其中每个阶段用 stage 关键字表示，而每个阶段又可以由多个 steps 组成。

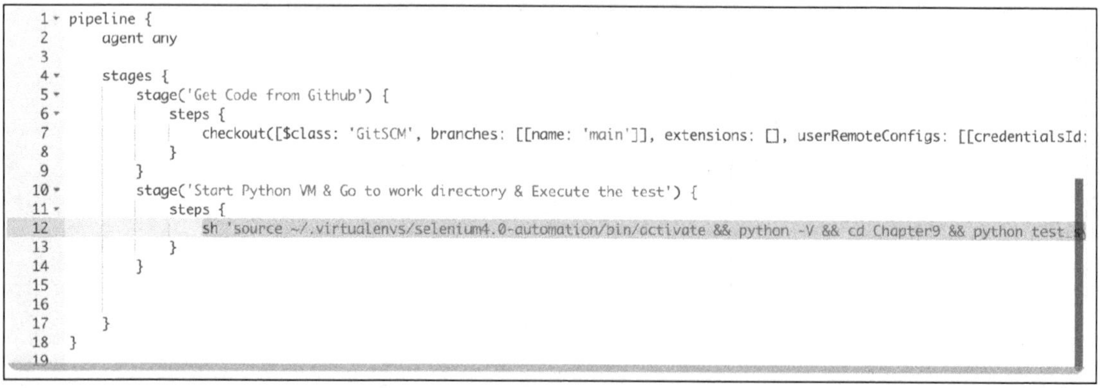

图 12.56

Pipeline 创建完毕后，便可以执行，具体如图 12.57 所示。单击"Build Now"，在 Pipeline 执行过程中，会在本地打开 Chrome 浏览器，如果执行过程没有异常产生，会正确登录项目系统。

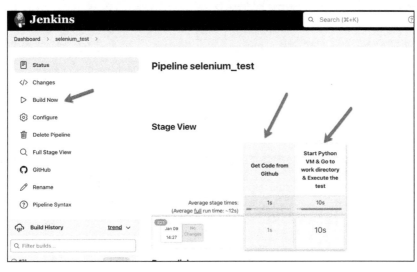

图 12.57

执行 Pipeline 后，在当前执行测试的目录下，创建项目登录成功后的浏览器页面截图文件，文件名为"login_screenshot.png"，截图如图 12.58 所示。

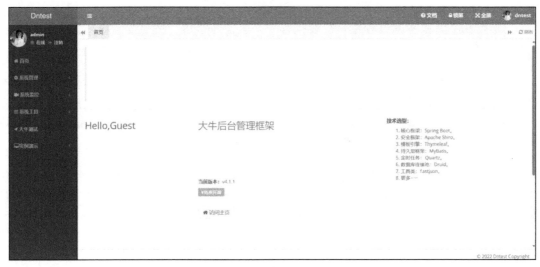

图 12.58

12.5 完整的 Jenkins 自动化实例

通过前面的介绍，我们对 Git、Jenkins 等相关内容有了大致的了解。本节将串联各个知识点，以一个综合性的案例来更加深入地讲解与 Jenkins 相关的内容。

以设置网易邮箱为例，先开启网易邮箱的 SMTP 服务，在网易邮箱首页单击"设置"，在下拉菜单中选择"POP3/SMTP/IMAP"，单击"开启"按钮即可开启 POP3/SMTP/IMAP 服务，具体如图 12.59 所示。

图 12.59

在开启 SMTP 服务的过程中，需要执行安全验证等步骤，按照提示操作即可，开启后会产生一个 16 位授权码，保存好这个授权码，稍后在 Jenkins 配置中会用到。

Jenkins 中的邮件设置，选择"Manage Jenkins → Configure System → E-mail Notification"，进入如图 12.60 所示界面，在"SMTP server"下输入"smtp.163.com"，勾选"Use SMTP Authentication"选项，在"User Name"下输入网易邮箱地址，在"Password"下输入通过上一步得到的 16 位授权码，在"SMTP Port"下输入"25"。

在"Extended E-mail Notification"区域，如图 12.61 所示，在"SMTP server"下输入"smtp.163.com"，在"SMTP Port"下输入"25"，在"Credentials"下选择登录凭证，配置 SMTP 的邮箱和 16 位授权码，在"Default Content Type"下一般选择"HTML(text/html)"。

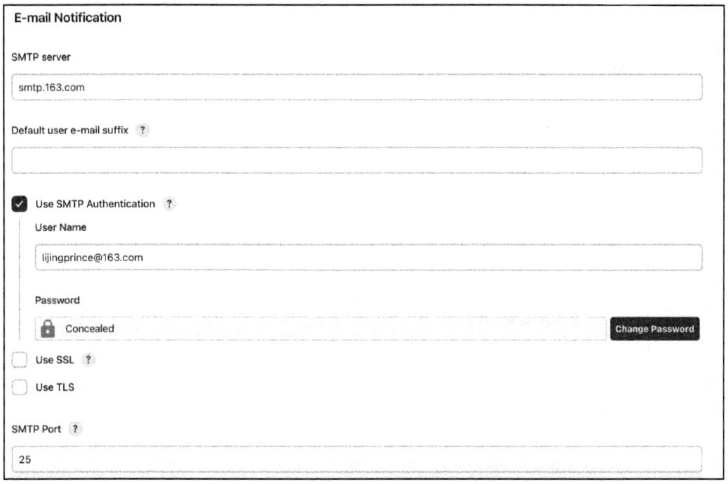
E-mail Notification
SMTP server
smtp.163.com
Default user e-mail suffix
Use SMTP Authentication
User Name
lijingprince@163.com
Password
Concealed
Change Password
Use SSL
Use TLS
SMTP Port
25

图 12.60

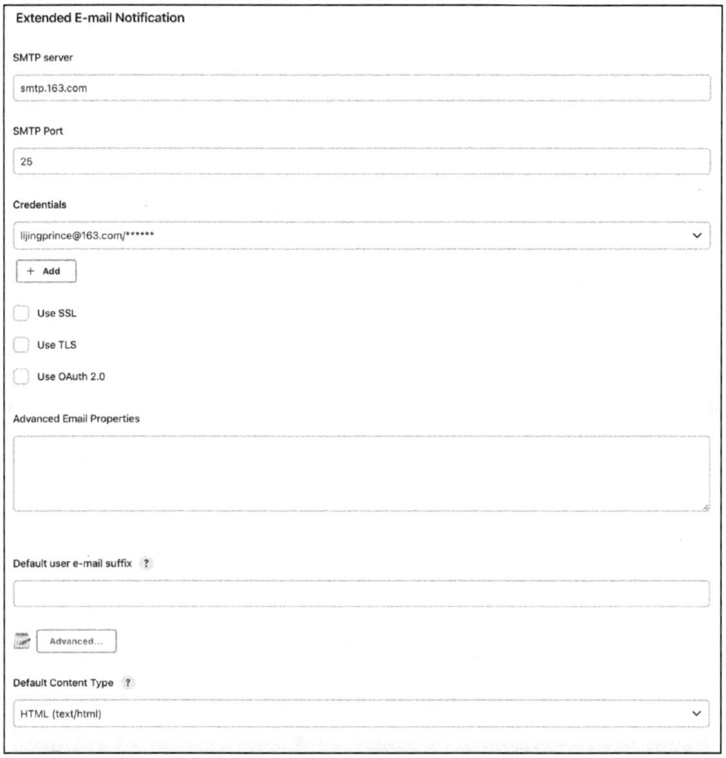
Extended E-mail Notification
SMTP server
smtp.163.com
SMTP Port
25
Credentials
lijingprince@163.com/******
+ Add
Use SSL
Use TLS
Use OAuth 2.0
Advanced Email Properties
Default user e-mail suffix
Advanced...
Default Content Type
HTML (text/html)

图 12.61

在该区域中还需要设置 Default Triggers（默认触发条件），可以根据自己的需要进行设置，具体如图 12.62 所示，如果勾选 "Always" 选项，每次构建项目时都会发邮件。

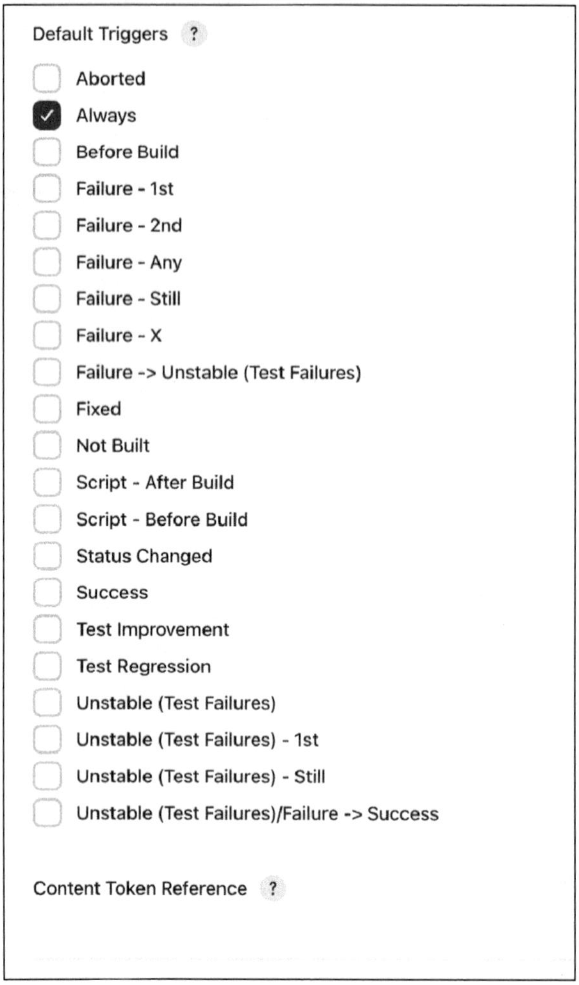

图 12.62

如何测试邮箱设置是否正常呢？可以在 "Configure System" 界面，向下滚动到页面最下面，具体如图 12.63 所示。在 "Test e-mail recipient" 下输入接收邮件的邮箱地址，然后单击 "Test configuration" 按钮。

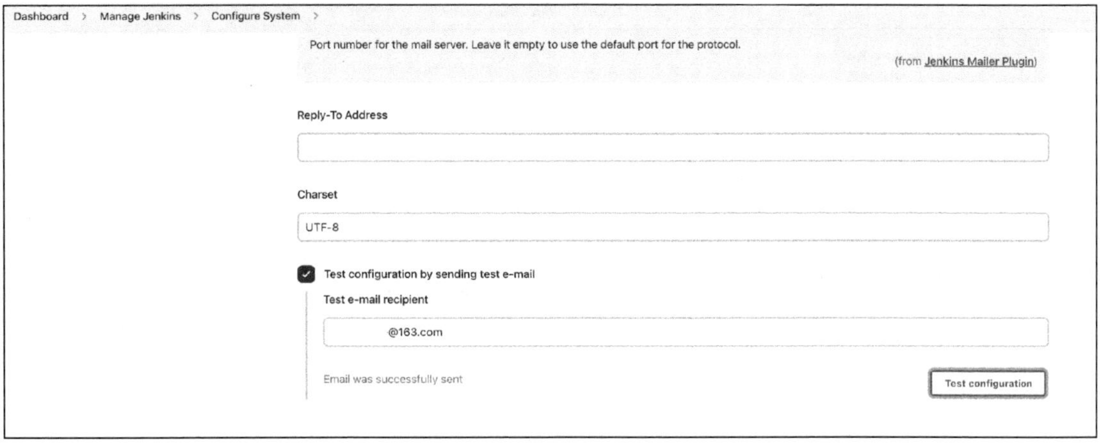

图 12.63

检查邮箱是否收到测试邮件。收到如图 12.64 所示的邮件，则表明 Jenkins 邮箱配置正确。

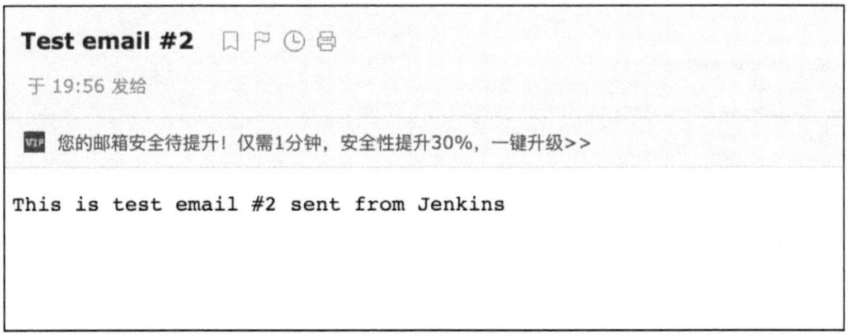

图 12.64

在 Jenkins 中新建一个自由风格项目，首先设置代码管理，如图 12.65 所示，需要设置代码仓库地址（Repository URL）、凭证（Credentials）、要构建的代码分支（Branches to build），等等。

设置构建触发方式，如图 12.66 所示。有如下两种常见触发方式，一般在实际应用中选择其中一种即可。

（1）定期构建项目（Build periodically），周期性地构建项目，不管项目代码有无更新都进行构建。如配置为"０ ９ ＊ ＊ ＊"，表明中国时间每天上午 9:00 必须构建一次代码。

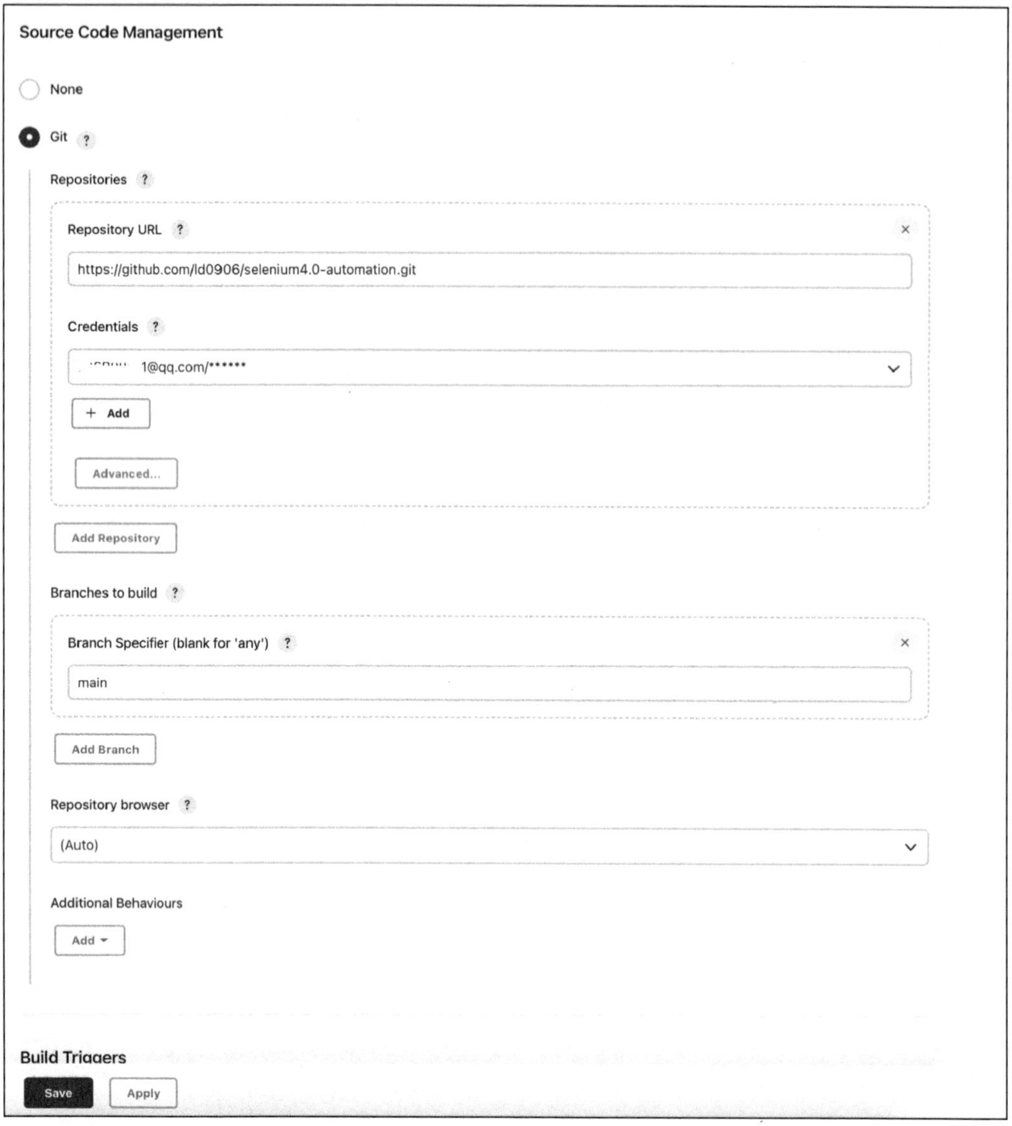

图 12.65

（2）使用 Poll SCM 模式构建项目，定时检查代码变更，如果有变更就拉取最新的代码，然后执行构建项目操作。如配置为"H/15 * * * *"，表明每 15 分钟检查一次代码变更。

第 12 章　测试平台维护与项目部署

图 12.66

下一步配置构建细节，如果 Jenkins 运行于 macOS 或 Linux 系统中，则可以选择"Execute shell"，在"Command"输入框中输入命令，具体如图 12.67 所示。

（1）开启 Python 虚拟环境（需要根据自己的实际情况，更改为准确的虚拟环境路径），命令为"source ~/.virtualenvs/selenium4.0-automation/bin/activate"。

（2）切换到代码目录，命令为"cd chapter10"。

（3）删除存在的报表文件，命令为"rm report_add_position.html"。

（4）执行测试，命令为"python test_system.py"。

如果 Jenkins 运行在 Windows 系统中，则可以选择"Execute Windows batch command"。在"Post-build Actions"中添加"Publish HTML reports"，如图 12.68 所示，需要安装插件，具体如图 12.69 所示，脚本执行后会生成一个 HTML 格式的测试结果，文件名为"report_add_position.html"。

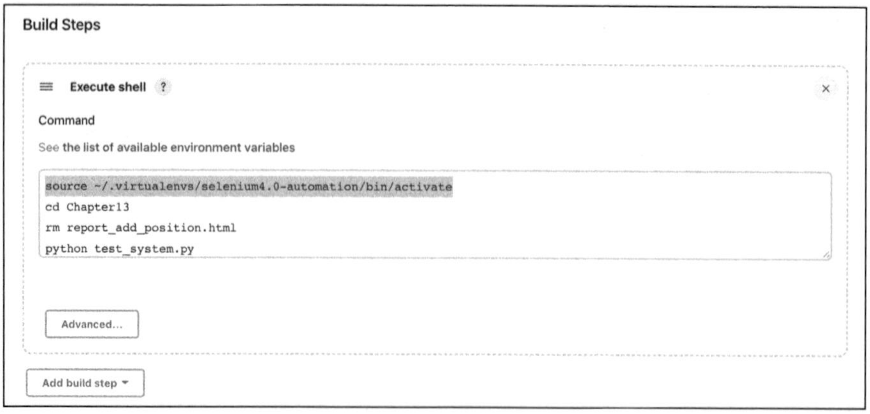

图 12.67

图 12.68

在"Post-build Actions"中添加发送邮件模块,具体如图 12.70 所示。在"Project Recipient List"下输入接收邮件的邮箱地址,不同邮箱地址之间用逗号隔开;在"Default Subject"和"Default Content"下定制化邮件标题和内容;在"Attachments"下设置邮件附件;在"Attach Build Log"下选择在邮件中将本次构建日志作为附件发送。

第 12 章 测试平台维护与项目部署

图 12.69

图 12.70

测试 Job 的执行结果如图 12.71 所示，测试报告的位置在如图 12.72 所示的位置。

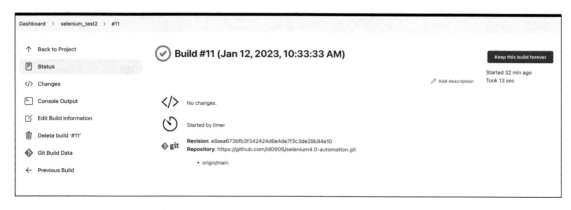

图 12.71

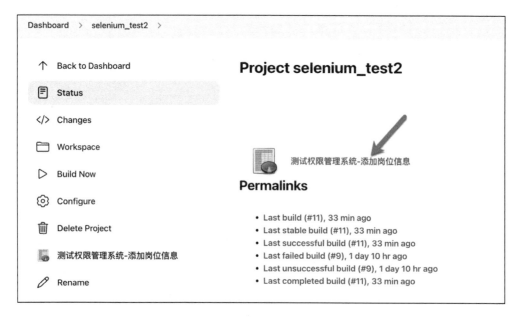

图 12.72

在项目构建完成之后，可以查看设置的邮箱是否能收到测试结果邮件，如图 12.73 所示，说明邮箱设置正常。

第 12 章 测试平台维护与项目部署

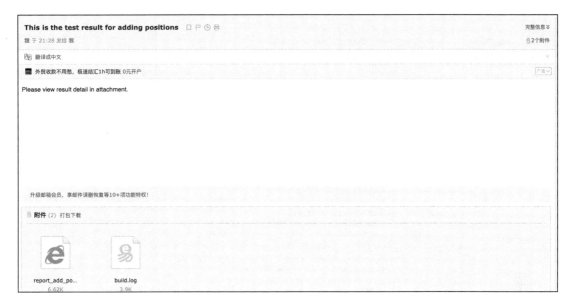

图 12.73

12.6 项目部署

在实际项目中，会遇到需要将项目部署到新环境的需求。对于 Selenium 来说，比较重要的是能完整地将要使用的功能模块安装到新环境中。在本节中，笔者将主要解决这个需求。

12.6.1 获取当前环境模块列表

获取当前环境模块列表的方法比较简单，在命令行中执行命令"pip freeze > requirements.txt"即可，执行结果如图 12.74 所示。

```
timdeMacBook-Pro:~ tim$ cat requirements.txt
asgiref==3.6.0
async-generator==1.10
attrs==22.2.0
behave==1.2.6
certifi==2022.12.7
charset-normalizer==3.0.1
ddt==1.6.0
Django==4.1.7
h11==0.14.0
idna==3.4
iniconfig==2.0.0
loguru==0.6.0
outcome==1.2.0
packaging==23.0
parse==1.19.0
parse-type==0.6.0
pluggy==1.0.0
py==1.11.0
PySocks==1.7.1
pytest==7.2.1
pytest-assume==2.4.3
pytest-html==3.2.0
pytest-metadata==2.0.4
PyYAML==6.0
requests==2.28.2
selenium==4.7.2
timdeMacBook-Pro:~ tim$ cat requirements.txt
asgiref==3.6.0
async-generator==1.10
attrs==22.2.0
behave==1.2.6
certifi==2022.12.7
charset-normalizer==3.0.1
ddt==1.6.0
Django==4.1.7
h11==0.14.0
idna==3.4
iniconfig==2.0.0
loguru==0.6.0
outcome==1.2.0
packaging==23.0
parse==1.19.0
parse-type==0.6.0
pluggy==1.0.0
py==1.11.0
PySocks==1.7.1
pytest==7.2.1
pytest-assume==2.4.3
pytest-html==3.2.0
pytest-metadata==2.0.4
PyYAML==6.0
requests==2.28.2
selenium==4.7.2
six==1.16.0
sniffio==1.3.0
sortedcontainers==2.4.0
sqlparse==0.4.3
tornado==6.2
trio==0.22.0
trio-websocket==0.9.2
urllib3==1.26.13
wsproto==1.2.0
xlrd==2.0.1
xlwt==1.3.0
```

图 12.74

12.6.2 安装项目移植所需的模块

从 12.6.1 节得到了项目移植所需的 Python 模块,所需的模块都列入了文件 requirements.txt 中,安装这些模块时需要执行命令"pip install -r requirements.txt"。

第 13 章 Docker 容器技术与多线程测试

在自动化测试项目中,测试环境可能有很多种,比如某个 Web 项目就需要在 Windows+IE 测试环境中进行测试或者在 Windows+Chrome 测试环境中进行测试等。全部按照传统方式部署、管理测试环境,会出现效率低的问题。用 Docker 技术搭建分布式环境,可以提高环境部署的效率,这也是自动化运维发展的趋势之一。在本章中,笔者将带领大家学习 Docker、多线程等知识。

13.1 Docker 简介

Docker 是基于 Go 语言实现的一个云开源项目,托管在 GitHub 上,任何人都可以参与。Docker 提供了一个轻量级的操作系统虚拟化解决方案,Docker 是一个开源引擎,可以创建轻量化、可移植的容器。下面介绍一下 Docker 的三大核心概念:镜像、容器和仓库。

（1）镜像类似于虚拟机的镜像，也类似于通常介绍的安装文件。

（2）容器类似于一个轻量级的沙箱，通过镜像创建应用运行实例，这些容器在 Docker 中被启动、停止和删除，并且这些容器是相互隔离、互相不可见的。

（3）仓库有些类似于代码仓库，是 Docker 集中存放镜像文件的场所。

Docker 的技术架构大致如图 13.1 所示。

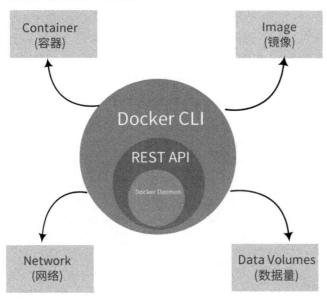

图 13.1

图 13.1 中各部分解释如下。

（1）最核心的部分是 Docker Daemon，也被称作 Docker 守护进程。它行使 Docker 服务端的职能，既可以部署在远端，也可以部署在本地。

（2）REST API，实现了客户端和服务端的交互协议，在通信之前开启相关服务即可。

（3）Docker CLI，CLI 是 Command Line Interface 的缩写，实现了利用 Docker 对容器和镜像进行管理的目的，并且为用户提供了统一的操作界面。客户端提供镜像，镜像可以创建一个或者多个容器（Container）。容器在 Docker 客户端只是一个进程，因为在实际应用中，打包好镜像，通过镜像来创建容器，在容器中运行应用即可。而服务端负责管理网络和磁盘，我们不用关心。

（4）Image，即镜像，可以自己创建镜像，也可以从网上下载，供自己使用。镜像包含了一个 RFS（根文件系统），在一个镜像中可以创建多个容器。

（5）Container 是由 Docker 客户端通过镜像创建的实例，用户运行的应用位于容器中，一旦容器实例创建成功，就可以被当作一个简单的根文件系统。每个应用运行在隔离的容器中，拥有独立的网络、权限和用户资源等。Docker 的这些机制可以确保容器的安全和互不影响等。

如图 13.2 所示，Docker 容器与虚拟机比较，最大的区别是容器之间会共享主机内核和操作系统，它们分别运行着独立的进程，相比之下虚拟机是和主机隔离的另一套操作系统，能够通过 Hypervisor 技术访问主机资源。图中，CONTAINER 表示容器，VM 表示虚拟机，App A、App B 和 App C 表示容器或虚拟机内的应用程序，Bins/Libs 表示容器或虚拟机内的可执行程序或库文件，Guest OS 是虚拟机内独立于主机的操作系统，Host OS 表示主机操作系统，Infrastructure 表示容器或虚拟机都需要的底层基础架构。

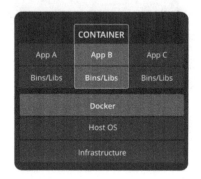 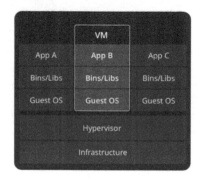

图 13.2

13.2　Docker 的一般应用场景

13.1 节讲解了与 Docker 技术相关的基础知识。接下来介绍一下 Docker 常见的应用场景，有以下几种。

（1）项目打包部署：传统的部署手段是先安装/部署一大堆依赖的工具、软件等，在这种情况下出现错误的概率极大，导致了效率低下。而 Docker 可以实现类似环境打包的概念，可以先将环境打包到镜像，再直接根据镜像启动容器，这样提高了部署效率，也降低了出现错误的概率。

(2) Web 应用的自动化部署和冒烟测试验证。

(3) 运行自动化测试和项目持续集成、发布。

(4) 在微服务环境中部署和调整数据库或其他的后台应用。

(5) 基于云平台的应用。

Docker 容器技术可以为企业解决的问题包括以下几种。

(1) 难以维护多种开发语言、多种运行环境。一个企业发展到一定规模的时候,有多种产品共存,涉及多种开发语言、多种运行环境。一个项目可能就有一种运行环境,每次需要开发或者测试时,就需要重新部署一遍,再进行开发或者测试。这种传统的模式会产生一定的问题,比如可能会导致每次开发或者测试的结果都不一样,还有就是浪费了时间。而如果使用容器技术,则可以创建镜像,只需要管理镜像,当需要的时候直接拉取镜像即可。

(2) 环境不一致引发的困惑。这种场景在很多公司都会出现,在引入容器技术之前,开发人员有自己的本地开发环境,测试人员有测试环境,此外还有生产环境或预生产环境。比如开发人员完成了某个功能的开发工作,在本地测试通过后提交给测试人员做软件测试,而测试人员发现了问题,这种问题很可能是由于环境不一致造成的。如果引入了 Docker 容器技术,那就不一样了,在完成了开发工作后,会创建镜像到镜像库中,而测试环境、生产环境下要做的是,从镜像库中拉取镜像。容器技术提供了一种类似标准工具的作用。

(3) 微服务架构的挑战。将一个大的应用拆分成多个小的微服务,这对运维工程师等角色来说挑战性极大。容器技术在这种情况下能体现独特的优势。

13.3 Docker 的安装和简单测试

13.3.1 Docker 的安装

本书中的 Docker 部署在 64 位 Linux 系统上,操作系统内核版本高于 6.10。基本安装步骤如下。

(1) 用具有 root 权限的用户登录终端,执行安装操作。

(2) 检查内核版本的命令为 "uname -r",如图 13.3 所示。

```
[root@VM-4-4-centos ~]# uname -r
3.10.0-1160.71.1.el7.x86_64
[root@VM-4-4-centos ~]#
```

图 13.3

（3）更新 yum，确保是最新版本，更新命令为"yum update"。命令执行完毕后，如图 13.4 所示（部分截图），新安装了 3 个软件包，升级更新了 85 个软件包，总共下载包大小为 152MB。

```
sssd-client              x86_64      1.16.5-10.el7_9.15      updates   230 k
sudo                     x86_64      1.8.23-10.el7_9.3       updates   844 k
systemd                  x86_64      219-78.el7_9.7          updates   5.1 M
systemd-devel            x86_64      219-78.el7_9.7          updates   216 k
systemd-libs             x86_64      219-78.el7_9.7          updates   419 k
systemd-python           x86_64      219-78.el7_9.7          updates   146 k
systemd-sysv             x86_64      219-78.el7_9.7          updates    97 k
tuned                    noarch      2.11.0-12.el7_9         updates   270 k
tzdata                   noarch      2022g-1.el7             updates   490 k
unzip                    x86_64      6.0-24.el7_9            updates   172 k
util-linux               x86_64      2.23.2-65.el7_9.1       updates   2.0 M
virt-what                x86_64      1.18-4.el7_9.1          updates    30 k
xfsdump                  x86_64      3.1.7-2.el7_9           updates   308 k
xz                       x86_64      5.2.2-2.el7_9           updates   229 k
xz-devel                 x86_64      5.2.2-2.el7_9           updates    46 k
xz-libs                  x86_64      5.2.2-2.el7_9           updates   103 k

Transaction Summary
================================================================================
Install    3 Packages
Upgrade   85 Packages

Total download size: 152 M
```

图 13.4

（4）添加 yum 仓库，命令如下。在命令执行完毕后，结果如图 13.5 所示。

```
tee /etc/yum.repos.d/docker.repo <<-'EOF'
[dockerrepo]
name=Docker Repository
baseurl=https://***.dockerproject.org/repo/main/centos/$releasever/
enabled=1
gpgcheck=1
gpgkey=https://***.dockerproject.org/gpg
EOF
```

```
[root@VM-4-4-centos ~]# tee /etc/yum.repos.d/docker.repo <<-'EOF'
> [dockerrepo]
> name=Docker Repository
> baseurl=https://███.dockerproject.org/repo/main/centos/$releasever/
> enabled=1
> gpgcheck=1
> gpgkey=https://███.dockerproject.org/gpg
> EOF
[dockerrepo]
name=Docker Repository
baseurl=https://███.dockerproject.org/repo/main/centos/$releasever/
enabled=1
gpgcheck=1
gpgkey=https://███.dockerproject.org/gpg
[root@VM-4-4-centos ~]#
```

图 13.5

（5）安装 Docker 的命令为"yum install -y docker-engine"。

（6）启动 Docker 服务，命令为"systemctl start docker.service"。

（7）Docker 服务启动后，需要验证 Docker 安装是否有问题，用命令"docker version"进行验证。如图 13.6 所示，会有客户端和服务端两部分的信息，表明 Docker 服务启动成功。

```
Client:
 Cloud integration: v1.0.22
 Version:           20.10.12
 API version:       1.41
 Go version:        go1.16.12
 Git commit:        e91ed57
 Built:             Mon Dec 13 11:46:56 2021
 OS/Arch:           darwin/amd64
 Context:           default
 Experimental:      true

Server: Docker Desktop 4.5.0 (74594)
 Engine:
  Version:          20.10.12
  API version:      1.41 (minimum version 1.12)
  Go version:       go1.16.12
  Git commit:       459d0df
  Built:            Mon Dec 13 11:43:56 2021
  OS/Arch:          linux/amd64
  Experimental:     false
 containerd:
  Version:          1.4.12
  GitCommit:        7b11cfaabd73bb80907dd23182b9347b4245eb5d
 runc:
  Version:          1.0.2
  GitCommit:        v1.0.2-0-g52b36a2
 docker-init:
  Version:          0.19.0
  GitCommit:        de40ad0
```

图 13.6

（8）如果要使 Docker 服务在开机时自动启动，可使用命令"sudo systemctl enable docker"进行设置。在当今的市场上，一般服务器是常开的，不会经常被重启，如各种云服务器。至此，Docker 在 Linux 上的安装已经完成。

13.3.2　Docker 的简单测试

打开一个命令行窗口，用"hello-world"镜像来测试 Docker 的安装是否有效。执行命令"docker run hello-world",结果显示如图 13.7 所示,结果表明 Docker 在 macOS 本地的安装正确,并且列出了 Docker 创建"hello-world"容器的流程。

```
[sh-3.2# docker run hello-world
Unable to find image 'hello-world:latest' locally
latest: Pulling from library/hello-world
2db29710123e: Pull complete
Digest: sha256:6e8b6f026e0b9c419ea0fd02d3905dd0952ad1feea67543f525c73a0a790fefb
Status: Downloaded newer image for hello-world:latest

Hello from Docker!
This message shows that your installation appears to be working correctly.

To generate this message, Docker took the following steps:
 1. The Docker client contacted the Docker daemon.
 2. The Docker daemon pulled the "hello-world" image from the Docker Hub.
    (amd64)
 3. The Docker daemon created a new container from that image which runs the
    executable that produces the output you are currently reading.
 4. The Docker daemon streamed that output to the Docker client, which sent it
    to your terminal.

To try something more ambitious, you can run an Ubuntu container with:
 $ docker run -it ubuntu bash

Share images, automate workflows, and more with a free Docker ID:
 https://hub.██████.com/

For more examples and ideas, visit:
 https://docs.██████.com/get-started/
```

图 13.7

常用的 Docker 命令有以下几种。

(1) docker container ls:查看容器情况。

(2) docker container stop <容器名>:停止某个容器。

(3) docker container ls -a:列出所有的容器。

(4) docker container rm webserver:删除容器。

(5) docker image ls:查看镜像情况。

(6) docker image rm <此处输入镜像名>:删除镜像。

(7) docker exec -it <这里输入容器 ID> bash:进入一个容器。

13.4　Python 多线程介绍

多线程的概念是相对单线程而言的。所谓单线程是指 CPU 在处理完成一项任务之前是不会开始处理第二项任务的。随着科技的进步，CPU 等计算机组件日新月异，CPU 处理速度越来越快，可以并行处理多项任务。

在 Python 中，多线程相关的模块主要有 Thread、Threading 和 Queue。其中 Thread 是多线程的底层支持模块，一般不建议使用。Threading 模块对 Thread 模块进行了封装，以对象的方式操作线程。而 Queue 模块实现了多生产者（Producer）、多消费者（Consumer）的队列模式。本书中涉及的多线程，主要使用的是 Threading 模块。

13.4.1　一般方式实现多线程

在实际项目中，"一般方式实现多线程"用得不多，因为会出现代码管理问题，如代码松散和维护困难。该实现方式通过初始化 Thread 类，在初始化的同时传入 target 与 args 参数值，用 start 方法启动线程。多线程的 join 方法实现了等待当前线程执行完毕才开始执行后续线程，Python 官方文档中给出的解释是 "Wait until the thread terminates"。在使用 join 方法时，也可以设置 timeout 参数，以防线程一直不结束而影响主线程的执行。示例代码如下，执行结果如图 13.8 所示。

```
#coding=utf-8
import threading
from time import sleep
#定义多线程要使用的方法
def display_name(user_name):
    sleep(2)
    print('用户名为： %s' % user_name)
#主线程的测试代码，多线程的操作在这里实现
if __name__ == '__main__':
    t1 = threading.Thread(target=display_name, args=('小王',))
    t2 = threading.Thread(target=display_name, args=('小张',))
    t1.start()
    t2.start()
```

```
t1.join(1)
t2.join(1)
print('线程操作结束!')
```

```
(selenium4.0-automation) jason118@192 Chapter13 % python multi_threads.py
用户名为: 小王
用户名为: 小张
线程操作结束!
```

图 13.8

13.4.2 用可调用类作为参数实例化 Thread 类

与上一种方式最大的区别是，第二种方式声明了一个可调用类。在可调用类中定义了"__call__"方法，该方法是可调用类的核心方法，在类初始化时传入参数，从而实现多线程操作。示例代码如下，执行结果如图 13.9 所示。

```
#coding=utf-8
import threading
from time import sleep
def dispaly_name(user_name):
    sleep(2)
    print('用户名为: %s' % user_name)
class MyThread(object):
    def __init__(self,func,args,name=""):
        self.func = func
        self.args = args
        self.name = name
    def __call__(self):
        self.func(self.args)
if __name__ == "__main__":
    t1 = threading.Thread(target=MyThread(func=dispaly_name,args=("小王",)))
    t2 = threading.Thread(target=MyThread(func=dispaly_name,args=("小张",)))
    t1.start()
    t2.start()
    t1.join(1)
    t2.join(2)
    print('线程操作结束! ')
```

```
(selenium4.0-automation) jason118@192 Chapter13 % python thread_call.py
用户名为： 小王
用户名为： 小张
线程操作结束！
```

图 13.9

13.4.3 Thread 类派生子类（重写 run 方法）

通过派生子类实现多线程是比较简单的方式。多线程的类采用直接继承 threading.Thread 的方式来得到相应的方法和属性。示例代码如下，执行结果如图 13.10 所示。

```python
#coding=utf-8
import threading
from time import sleep
def dispaly_name(user_name):
    sleep(2)
    print('用户名为： %s' % user_name)
class MyThread(threading.Thread):
    def __init__(self,func,args):
        super(MyThread,self).__init__()
        self.func = func
        self.args = args
    def run(self):
        print("ThreadName: "+self.name)
        self.func(self.args)
if __name__ == "__main__":
    t1 = MyThread(func=dispaly_name,args=("小王",))
    t2 = MyThread(func=dispaly_name,args=("小张",))
    t1.start()
    t2.start()
    t1.join(1)
    t2.join(1)
    print("线程操作结束！")
```

```
(selenium4.0-automation) jason118@192 Chapter13 % python thread_subClass.py
ThreadName: Thread-1
ThreadName: Thread-2
用户名为： 小王
用户名为： 小张
线程操作结束！
```

图 13.10

13.5 使用 Docker 容器技术进行多线程测试

本节介绍如何将 Docker 容器技术与多线程技术结合起来进行测试。在正式开始前，需要先了解 Selenium Grid 的相关知识。

13.5.1 Selenium Grid

Grid 组件是 Selenium 非常重要的功能组件之一，它主要用于远程分布式测试或者多浏览器并发测试。

Grid 目前的最新版本是 4 版本，和前一个版本有很大的不同，重构了很多内容，也应用了很多新技术。

参考 Selenium 官网，Grid 内的不同组件之间的关系如图 13.11 所示。

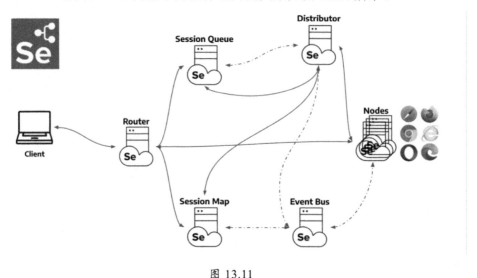

图 13.11

Grid 中一般包括 Router、Distributor、Session Map、Session Queue、Node 和 Event Bus 等组件。

Router 组件是 Grid 的入口，负责接收所有的外部请求，并且负责分发这些请求到对应的正确的其他组件进行处理。如果请求属于新的 Session 请求，会被分发给 Session Queue 组件；如果请求属于已经存在的 Session 请求，Router 会通过查询 Session Map 组件得到 Node 的 ID 信息，然后将请求直接分发给找到的 Node 进行处理。Router 还可以平衡 Grid 网络中的请求压力。

Distributor 组件有两个主要职责。

第一，处理 Node 注册请求并跟踪所有 Node 及其性能表现。

第二，轮询 Session Queue 中的请求并且处理挂起的新 Session 请求。

Session Map 组件，可以把它理解为一种 map 数据，该数据存储着 Session ID 和 Node 之间的关联信息。Session Map 为 Router 分发请求到某个 Node 提供了可能。

Session Queue 组件是一个先进先出的队列对象，总体来说，所有的新 Session 请求都会被 Router 组件分发给此队列。

Node 组件是 Grid 中的主角，是执行测试的主体。每个 Node 管理着运行它的机器上可用的浏览器信息。一个 Node 在可用之前，需要先通过 Event Bus 到 Distributor 组件进行注册。

Event Bus 组件充当着信道的角色，为其他组件之间进行通信提供信道，如 Node、Distributor、Session Queue 和 Session Map 等。

在 Grid 中选择角色，如果想单独启动每一个组件，选 Hub 和 Node 这样的角色；如果所有组件都需要运行在一台机器上，则选择 Standalone 运行模式。

至此我们对 Selenium Grid 有了基本的认识，重点需要掌握其工作原理、特性及各种组件之间的工作关系。了解这些基本内容之后，再来了解在 Docker 下安装 Selenium Grid 测试环境的细节。

13.5.2 安装需要的镜像

使用 Grid 在 Chrome 和 Firefox 浏览器上做自动化测试，有如下 3 个镜像需要下载安装。安装完镜像之后使用命令"docker images"查看镜像情况，如图 13.12 所示，镜像已经安装好。

- selenium/hub

- selenium/node-firefox

- selenium/node-chrome

```
jason118@192 ~ % docker images
REPOSITORY              TAG       IMAGE ID        CREATED         SIZE
selenium/node-firefox   latest    5e4a0d1488ac    2 months ago    1.17GB
selenium/node-chrome    latest    de92c5f8374c    2 months ago    1.3GB
selenium/hub            latest    cfb1e9a75197    2 months ago    420MB
```

图 13.12

13.5.3 启动 Selenium Hub

新版 Grid 在启动 Hub 时需要创建一个 Docker 网络，可以执行命令"docker network create grid"，然后执行 Hub 启动命令"docker run -d -p 4442-4444:4442-4444 --net grid --name selenium-hub selenium/hub"。其中的"run"用于运行一个镜像，创建一个容器。命令执行完毕之后，查看容器启动情况，输入命令"docker ps"即可，运行结果如图 13.13 所示。

```
jason118@192 ~ % docker ps
CONTAINER ID   IMAGE           COMMAND                 CREATED          STATUS         PORTS                                              NAMES
44ad4b28be33   selenium/hub    "/opt/bin/entry_poin…"  6 seconds ago    Up 5 seconds   0.0.0.0:4442-4444->4442-4444/tcp                   selenium-hub
```

图 13.13

13.5.4 启动 Selenium Node

启动两个 Node 用于测试，只需要分别执行如下两个命令即可。命令执行成功后，运行命令"docker ps"查看当前的容器启停状态，如图 13.14 所示，说明 3 个容器都已经启动成功。如果要查看所有的容器状态（包括处于关闭状态的容器），使用命令"docker ps -a"。

```
jason118@192 ~ % docker ps
CONTAINER ID   IMAGE                    COMMAND                 CREATED          STATUS          PORTS                                              NAMES
ecf15786e8ac   selenium/node-firefox    "/opt/bin/entry_poin…"  22 seconds ago   Up 21 seconds   5900/tcp                                           pedantic_bose
332a69792889   selenium/node-chrome     "/opt/bin/entry_poin…"  31 seconds ago   Up 30 seconds   5900/tcp                                           compassionate_carson
44ad4b28be33   selenium/hub             "/opt/bin/entry_poin…"  2 minutes ago    Up 2 minutes    0.0.0.0:4442-4444->4442-4444/tcp                   selenium-hub
```

图 13.14

安装 chrome 节点的命令如下：

```
docker run -d --net grid -e SE_EVENT_BUS_HOST=selenium-hub \
```

```
    --shm-size="2g" \
    -e SE_EVENT_BUS_PUBLISH_PORT=4442 \
    -e SE_EVENT_BUS_SUBSCRIBE_PORT=4443 \
    selenium/node-chrome
```

安装 firefox 节点的命令如下：

```
docker run -d --net grid -e SE_EVENT_BUS_HOST=selenium-hub \
    --shm-size="2g" \
    -e SE_EVENT_BUS_PUBLISH_PORT=4442 \
    -e SE_EVENT_BUS_SUBSCRIBE_PORT=4443 \
    selenium/node-firefox
```

至此，一个 Hub 和两个 Node，从下载镜像到启动容器都已经结束，并且都能成功运行。

13.5.5 查看 Selenium Grid Console 界面

以上容器的安装准备工作已就绪，下一步需要在安装 Hub 的机器上查看 Selenium Grid 的配置情况。根据配置，网址 http://localhost:4444/ui/index.html#/界面如图 13.15 所示，从该图可知，Grid 的配置情况与我们设置安装的一致，图上也显示了配置的浏览器 Node 的相关细节，如浏览器的具体版本等。通过命令"docker logs selenium-hub"可以查看 Hub 下有多少 Node，结果如图 13.16 所示。

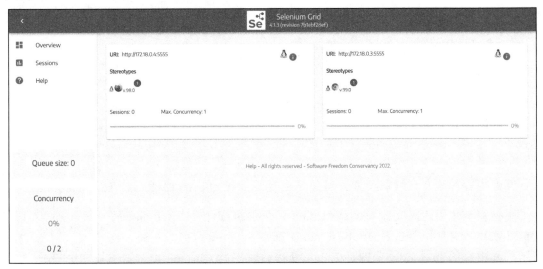

图 13.15

```
jason1180192:~ N docker logs selenium-hub
2023-02-18 18:03:28,768 INFO Included extra file "/etc/supervisor/conf.d/selenium-grid-hub.conf" during parsing
2023-02-18 18:03:28,774 INFO RPC interface 'supervisor' initialized
2023-02-18 18:03:28,774 CRIT Server 'unix_http_server' running without any HTTP authentication checking
2023-02-18 18:03:28,775 INFO supervisord started with pid 9
2023-02-18 18:03:29,781 INFO spawned: 'selenium-grid-hub' with pid 11
Tracing is disabled
2023-02-18 18:03:29,791 INFO success: selenium-grid-hub entered RUNNING state, process has stayed up for > than 0 seconds (startsecs)
18:03:30.687 INFO [LoggingOptions.configureLogEncoding] - Using the system default encoding
18:03:30.595 INFO [OpenTelemetryTracer.createTracer] - Using OpenTelemetry for tracing
18:03:30.766 INFO [BoundZmqEventBus.<init>] - XPUB binding to [binding to tcp://*:4442, advertising as tcp://172.18.0.2:4442], XSUB binding to [binding to tcp://*:4443, advertising as tcp://172.18.0.2:4443]
18:03:30.860 INFO [UnboundZmqEventBus.<init>] - Connecting to tcp://172.18.0.2:4442 and tcp://172.18.0.2:4443
18:03:30.901 INFO [UnboundZmqEventBus.<init>] - Sockets created
18:03:31.908 INFO [UnboundZmqEventBus.<init>] - Event bus ready
18:03:33.082 INFO [Hub.execute] - Started Selenium Hub 4.7.2 (revision 4d4020c3b7): http://172.18.0.2:4444
18:03:42.885 INFO [Node.<init>] - Binding additional locator mechanisms: id, name, relative
18:05:43.574 INFO [GridModel.setAvailability] - Switching Node 9d8e1526-2653-45f3-b23c-2431e8eb3d17 (uri: http://172.18.0.3:5555) from DOWN to UP
18:05:43.578 INFO [LocalDistributor.add] - Added node 9d8e1526-2653-45f3-b23c-2431e8eb3d17 at http://172.18.0.3:5555. Health check every 120s
18:05:51.725 INFO [Node.<init>] - Binding additional locator mechanisms: id, relative, name
18:05:51.889 INFO [GridModel.setAvailability] - Switching Node e373eddd-9473-4f5d-88e3-866f057d2fe4 (uri: http://172.18.0.4:5555) from DOWN to UP
18:05:51.890 INFO [LocalDistributor.add] - Added node e373eddd-9473-4f5d-88e3-866f057d2fe4 at http://172.18.0.4:5555. Health check every 120s
```

图 13.16

13.5.6 Docker 环境下多线程并发执行 Selenium Grid 测试

Docker 的 Selenium Grid 测试环境准备完毕后,执行第 9 章中优化过的测试代码,分别在 Chrome 和 Firefox 浏览器上做自动化测试。

首先界定测试范围,测试场景只选择登录项目系统,本节中使用的测试代码对第 9 章中优化后的测试代码进行了一些调整,具体如下。

文件 functions.py 的内容如下:

```
from selenium import webdriver
'''
函数 return_driver 用于返回 driver 对象
'''
def return_driver(browser_type):
    browser_des = None
    if browser_type == "firefox":
        browser_des = webdriver.DesiredCapabilities.FIREFOX.copy()
    elif browser_type == "chrome":
        browser_des = webdriver.DesiredCapabilities.CHROME.copy()
    browser_des["platform"] = "LINUX"
    return webdriver.Remote('http://localhost:4444/wd/hub',desired_capabilities=browser_des)
```

其余内容和第 9 章代码一样,可参见配套资源包 chapter13/docker_run_selenium。

运行测试代码,在文件 test_system.py 目录下执行命令 "python test_system.py"。访问 http://localhost:4444/ui/index.html#/sessions 查看 Selenium Grid 的 Session 信息,如图 13.17 所示,其中有两个 Node,分别是 Linux 操作系统下的 Chrome 浏览器节点和 Firefox 浏览器节点。

第 13 章　Docker 容器技术与多线程测试

图 13.17

查看某个 Session 的运行结果，可以单击 Session ID 旁边的摄像头图标，具体如图 13.18 所示。

图 13.18

在 Session 页面中，首先需要输入 LiveView 的密码，默认密码是 secret，具体如图 13.19 所示。

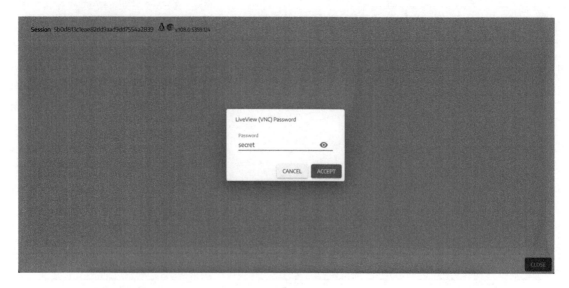

图 13.19

验证测试结果,即测试用户能否正确登录系统,通过查看两个 Session 的最终测试结果,可以发现:在 Chrome 与 FireFox 浏览器中,用户都可以正确登录系统。